T0331026

Engineering Dimensions, Units, and Conversions

Engineering Dimensions, Units, and Conversions delves into the analysis and application of the dimensions, units, and unit conversions in engineering practical use. It demonstrates the importance of dimensional homogeneity and unit consistency.

Offering a comprehensive exploration of both primary and secondary units, the book presents detailed portrayals of various unit systems in both the English system and the International System (SI). It provides insight into conversion ratios and introduces software-based methodologies. The book also examines dimensioning in drawings, including dimensioning basics and numerous exercises of object and system dimensioning.

The book will be a valuable reference for practicing engineers and researchers engaged in engineering research and development. It will also be of interest to undergraduate and graduate students majoring in engineering disciplines.

Engineering Dimensions, Units, and Conversions

Yongjian Gu

CRC Press
Taylor & Francis Group
Boca Raton London New York

CRC Press is an imprint of the
Taylor & Francis Group, an **informa** business

Designed cover image: Yongjian Gu

First edition published 2025
by CRC Press
2385 NW Executive Center Drive, Suite 320, Boca Raton FL 33431

and by CRC Press
4 Park Square, Milton Park, Abingdon, Oxon, OX14 4RN

CRC Press is an imprint of Taylor & Francis Group, LLC

ISBN: 978-1-032-83363-7 (hbk)
ISBN: 978-1-032-83362-0 (pbk)
ISBN: 978-1-003-50897-7 (ebk)

DOI: 10.1201/9781003508977

Typeset in Times
by Apex CoVantage, LLC

Contents

Preface

This book delves into the description, analysis, and application of the engineering dimensions, units, and unit conversions. Dimensions represent measurement types of physical quantities, while units denote the magnitudes assigned to the correlated measurements of these quantities. Both dimensions and units play a pivotal role in engineering practice. Failing to correctly apply them leads to confusion in calculations and drawings for engineering design and research and design (R&D).

Dimensions and units are categorized into the fundamental, also known as the primary, dimensions and units and the derived, also known as the secondary, dimensions and units. The fundamental dimensions and units are basics, while the derived dimensions and units are derived from the fundamental and can be expressed in terms of the fundamental ones. Proper unit expression in calculations and drawings follows conventional formats and rules, such as dimensioning methods, significant figures, and rounding off numbers. In engineering practice, the consistency of dimensions and units of quantities is crucial in physical expressions and equation calculations. Inconsistent dimensions and units should not be manipulated in equations, such as addition, subtraction, multiplication, and division.

In the United States, two unit systems are used in engineering practice. One is the English system also called the United States Customary System (USCS), and another is the International System (SI) also called the metric system. Unit conversions in the system and between the systems are essential to maintain unit consistency. Dimensional analysis is useful to help check the unit consistency in physical expressions and equation calculations and determine the relationship of dimensions between physical quantities. The dimensionless numbers resulting from dimensional analysis are extremely valuable to reveal the characteristics and behaviors of the complex phenomena in R&D.

This book is composed of eight chapters to cover the previously mentioned delineations:

Chapter 1 lays the groundwork by emphasizing the necessary application of dimensions and units in engineering and elucidating their indispensable role in engineering practice. It is so important for students majoring in engineering, engineers engaging in engineering practice, and those preparing for the FE/PE exams to thoroughly understand and correctly apply the dimensions units, and unit conversions.

Chapter 2 delves into the intricacies of dimensions in engineering, comprehensively describes the fundamental and derived dimensions, and illustrates the principle of dimensional homogeneity as a key factor in maintaining dimensional consistency. The principle of dimensional homogeneity is a basic axiom in dimensional analysis.

Chapter 3 ventures into the expansive landscape of units in engineering, offers a complete exploration of both fundamental and derived units, and portrays in detail various unit systems in both the English system and the SI. In the chapter, the guidelines of applying units in the SI, including common rules, decimal separators, and unit prefixes, are presented.

Chapter 4 is dedicated to the intricate process of the unit conversions, providing insights into the unit conversion factor and the unity conversion ratio and fully illustrating the unit conversions in the English system, the SI, and between the two systems. In the chapter, the unit conversion using computer software is also introduced.

Chapter 5 unveils the world of dimensional analysis, describes the methods of nondimensionalized equations and Pi theorem to obtain dimensionless numbers, and detailly presents the procedures to generate the dimensionless numbers, which help with understanding complex physical phenomena. In the chapter, commonly used dimensionless numbers are presented.

Chapter 6 takes a closer look at the role of units and unit conversions in equation calculations and underscores the importance of unit consistency. The chapter also delves into conventional rules for how to formulate numerical values in equation calculations, such as significant figures and rounding off numbers.

Chapter 7 provides an extensive discussion of basics of dimensioning objects in engineering drawings. Numerous illustrations of dimensioning objects on drawings are presented.

Chapter 8 offers basic and practical knowledge of pipes, ducts, and dimensioning pipes and ducts of pipe and duct systems. Typical drawings of dimensioning the pipe and duct systems are presented.

The book, enriched with numerous examples, exercises including some in the style found on the FE exam style, and drawings, serves as a comprehensive resource for students majoring in engineering, engineers engaging in engineering practice, and those preparing for the FE/PE exams to thoroughly understand and correctly apply the dimensions, units, and unit conversions. The book can also be used as a textbook in engineering colleges and a reference for engineers and researchers in engineering design and R&D.

Nomenclature

a	Acceleration	m/s², ft/s²
	Speed of sound	m/s, ft/s
A	Area	m², ft²
	Surface area	m², ft²
A_c	Flow cross sectional area	m², ft²
c	Speed of sound	m/s, ft/s
C	Chord line	m, ft
	Heat capacity	kg·m²/s²·K, blm·ft²/s²·R
C_f	Drag coefficient	—
	Fanning friction factor	—
	Force coefficient	—
	Friction coefficient	—
C_M	Moment coefficient	—
d	Diameter	m, ft
D	Diameter	m, ft
E	Energy	J or kJ, Btu or lbf·ft
	Material modulus of elasticity	kg/m·s², lbm/ft·s²
Eu	Euler number	—
f	Frequency	1/s
	Friction	N, lbf
F	Force	N, lbf
Fr	Froude number	—
g	Gram	—
	Gravitational acceleration	m/s², ft/s²
G	Universal constant of gravitation	N·m²/kg², lbf·ft²/lbm²
g_c	Gravitational constant	—, lbm·ft/lbf·s²
Gr	Grashof number	—
h	Height	m, ft
	Altitude	m, ft
	Elevation	m, ft
	Specific enthalpy	J/kg or kJ/kg, Btu/lbm
H	Height	m, ft
I	Electric current	amp, amp
	Area moment of inertia	m⁴, ft⁴
J	Mechanical equivalent of heat	—, 778 ft·lbf/Btu
	Luminous intensity	cd, cd
k	Thermal conductivity	W/m·K, Btu/s·ft·R
	Specific heat ratio	—
ke	Specific kinetic energy	J/kg or kJ/kg, Btu/lbm or lbf·ft/lbm
KE	Kinetic energy	J or kJ, Btu or lbf·ft
	Kinetic energy rate	J/s or kJ/s, Btu/s or lbf·ft/s
Knot	One nautical mile per hour	0.514 m/s, 1.151 mph

l	Length	m, ft
L	Liter	m³, —
L	Length	m, ft
m	Mass	kg, lbm
$\dot{m}$	Mass flow rate	kg/s, lbm/s
M	Mass	kg, lbm
	Molar mass	kg/kmol, lbm/lbmol
Ma	Mach number	—
N	Newton	kg·m/s², —
	Numerical value	—
N	Number of moles	—
p	Pressure	kPa or N/m², Psi or lbf/ft²
P	Power	W or kW, hp
	Pressure	kPa or N/m², Psi or lbf/ft²
	Work	J or kJ, Btu or lbf·ft
P_a	Absolute pressure	kPa or N/m², Psi or lbf/ft²
pe	Specific potential energy	J/kg or kJ/kg, Btu/lbm or lbf·ft/lbm
PE	Potential energy	J or kJ, Btu or lbf·ft
P_o	Atmospheric pressure	kPa or N/m², Psi or lbf/ft²
P_∞	Surrounding pressure	kPa or N/m², Psi or lbf/ft²
q	Unit of measurement	—
q	Heat exchange	kJ/kg, Btu/lbm
Q	Heat energy	kJ, Btu
	Physical quantity	—
$\dot{Q}$	Heat transfer rate	kW, Btu/h
r	Radius	m, ft
R	Gas constant	kJ/kg·K, Btu/lbm·R
	Radius	m, ft
Re	Reynolds number	—
R_u	Universal gas constant	8.31447 kJ/kmol·K, 1.9858Btu/kmol·R
SG	Specific gravity	—
St	Strouhal number	—
t	Time	s
T	Temperature,	°C or K, °F or R
	Time	s
	Torque	N·m, lbf·ft
	Total kinetic energy	N·m, ft·lbf
T_H	High temperature of source	°C or K, °F or R
T_L	Low temperature of sink	°C or K, °F or R
V	Velocity	m/s, ft/s
	Volume	m³, ft³
$\dot{V}$	Volume flow rate	m³/s, ft³/s
w	Specific work	J/kg or kJ/kg, Btu/lbm or lbf·ft/lbm
W	Work	J or kJ, Btu or lbf·ft
	Weight	N, lbf
$\dot{W}$	Power	W or kW, hp or Btu/h

| y_c | Critical depth | m, ft |
| z | Elevation | m, ft |

GREEK

α	Angular acceleration	rad/s^2
ε	Surface roughness	m, ft
ρ	Density	kg/m^3, lbm/ft^3
γ	Specific weight	N/m^3, lbf/ft^3
η_{th}	Thermal efficiency	%
μ	Viscosity, Absolute viscosity	kg/m·s, lbm/ft·s
v	Kinematic viscosity	m^2/s, ft^2/s
θ	Time	s
$\dot{\theta}$	Angular velocity	Rad/s
σ_s	Surface tension	m/s^2, ft/s^2
τ	Torque	N·m, ft·lbf
ω	Angular velocity	Rad/s

Abbreviations

ABET	Accreditation Board for Engineering and Technology
ANSI	American National Standards Institute
amp	ampere
ASHRAE	American Society of Heating, Refrigerating and Air-Conditioning Engineers
ASME	American Society of Mechanical Engineers
ASTM	American Society for Testing and Materials
avg	average
atm	atmosphere
BIPM	International Bureau of Weights and Measures (French: Bureau International des Poids et Mesures)
Btu	British thermal unit
CAD	computer-aided design
cd	candela
CFM	cubic feet per minute
CGPM	General Conference of Weights and Measures (French: *Conférence Générale des Poids et Mesures*)
CGS	centimeter–gram–second
CIPM	International Committee of Weights and Measures (French: Comité international des poids et mesures)
DN	diameter nominal
EAC	Engineering Accreditation Commission in ABET
FE	Fundamentals of Engineering Exam
FLθT	force–length–time–temperature
FMLθTQ	force-mass-length-time-temperature-heat
HVAC	heating, ventilation, and air conditioning
ID	internal diameter
IEC	International Electrochemical Commission
ISO	International Organization for Standardization (French: Organisation internationale de normalization)
kg	kilogram
kip	kilo pound-force
kn	knot
kN	kilonewton
kPa	kilopascal
kW	kilowatt
max	maximum
MKS	meter–kilogram–second
MLθT	mass–length–time–temperature
NCEES	National Council of Examiners for Engineers and Surveyors
NPS	nominal pipe size
N-S	Navier-Stokes

OD	out diameter
OTEC	ocean thermal energy conversion
PE	Principles and Practice of Engineering Exam
P.E.	licensed professional engineer
psi	pound-force per square inch
Pa	pascal
R&D	research and development
rpm	revolution per minute
SI	International System
	(French: Système International)
UCF	unit conversion factor
UCR	unity conversion ratio
USCS	United States Customary System
USCU	United States Customary Units

About the Author

Yongjian Gu received his M.Sc. and Ph.D. degrees in mechanical engineering from the State University of New York (SUNY) at Stony Brook. He is a professor at the U.S. Merchant Marine Academy (USMMA) and an adjunct at the New York Institute of Technology (NYIT). He teaches thermal fluids, total energy systems and design, propulsions, steam turbine and components, gas turbine and auxiliary equipment, engineering economics, and other courses to undergraduates and graduate students. He gives review and practice workshops of the Fundamentals of Engineering (FE) exam in the mechanical discipline. He holds a license as a registered Professional Engineer (P.E.) in the state of New York. He also holds a certificate as a professional Oracle Database Administrator (DBA). Prior to teaching in academic institutions, he worked at industrial corporations, consulting firms, and the U.S. Department of Energy's national laboratory as a senior mechanical engineer, a project engineer, and a lead engineer for many years. He has rich and valuable academic and professional experiences in HVAC system, piping systems, thermal energy systems, and renewable/sustainable energy applications. He is very knowledgeable about dimensions, units, and unit conversion in engineering. He is actively involved in engineering research and development and academic activities. He has many publications in his subject areas of teaching and professional expertise. He serves as a peer reviewer for multiple scientific and technical journals. He also sits on several professional boards and conference committees.

1 Introduction

1.1 QUANTITIES AND EQUATIONS IN ENGINEERING

Quantifies

A quantity in engineering, also known as a physical quantity, is a property of a material or a system that can be quantified by measurement. The quantity can be expressed as a combination of a numerical value and a unit of measurement. The numerical value is any real or integer number such as 4.851 or 72. If the quantity is denoted as a symbol, Q, the numerical value is N, and the unit of measurement is q, the relation of them is denoted as

$$Q = Nq \tag{1.1}$$

Typical physical quantities used in engineering are mass, length, pressure, velocity, force, and so on. For example, the light velocity c in a vacuum is 299,792,458 meters per second (m/s), in which 299,792,458 is the numerical value of c when the unit of measurement is of m/s, that is,

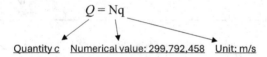

Quantity c Numerical value: 299,792,458   Unit: m/s

   In the preceding expression, the quantity of light velocity c in vacuum is 299,792,458 m/s, which must be a combination of the numerical value and its unit. The expression of $c = 299{,}792{,}458$ without the unit is typically called a pure number, which is confusing and meaningless in engineering applications because the value 299,792,458 can be in feet per second (ft/s), meters per hour (m/h), or millimeters per second (mm/s). If a different unit is used other than m/s, the numerical value will not be 299,792,458 in a vacuum condition. Table 1.1 shows some commonly used quantities and their symbols in engineering applications.

Equations

In general, there are two kinds of equations: the numerical-value equation and the quantity equation. In a numerical-value equation, the numerical values of the terms are pure numbers without needing to use units, for example,

$$72 + 56.4 = X, \qquad X = 128.4 \tag{1.2a}$$

and

$$36 \times 3.2 = Y, \qquad Y = 115.2 \tag{1.2b}$$

DOI: 10.1201/9781003508977-1

TABLE 1.1

Some Quantities and Their Symbols Commonly Used in Engineering

Quantity	Symbol
Acceleration	a
Area	A
Density	ρ
Energy	E
Force	F
Gravitational acceleration	g
Heat	Q
Length	l
Light velocity	c
Mass	m
Moment	M
Power	P
Pressure	p
Time	t
Temperature	T
Velocity	V
Volume	V
Weight	W
Work	W

Using the numerical-value equation such as Equations (1.2a) and (1.2b) in engineering practice, the equation is valid only when each numerical value is referenced to the same unit (assuming in the preceding equations, the units were omitted, but the omission is strongly not recommended) or when the equation has an implied physical meaning; for example, the displacement S of an article is determined by the product of particle moving speed V and the time span Δt, that is, $S = V \times \Delta t$, in which V and t have different units. In contrast, a quantity equation expresses a relation among quantities. The different values of the quantity Q can be obtained from the calculations with the different units; for example, the results using Equations (1.2a) and (1.2b) for the same quantity measurement but with different units can be

$$72 \text{ m} + 56.4 \text{ m} = X, \qquad X = 128.4 \text{ m} \qquad (1.3a)$$

or

$$239.98 \text{ ft} + 187.98 \text{ ft} = X, \qquad X = 427.96 \text{ ft}$$

and

$$36\frac{\text{m}}{\text{s}} \times 3.2 \text{ s} = Y, \qquad Y = 115.2 \text{m} \qquad (1.3b)$$

or

$$119.99\frac{ft}{s} \times 3.2\,s = Y, Y = 383.96\,ft$$

In the preceding equations, m, ft, m/s, ft/s, and s are the units of the quantities, respectively. In engineering applications, quantity equations are commonly used. Numerical-value equations and pure numbers should be avoided because they may cause confusion. *Exercise 1.1* and *Exercise 1.2* in Section 1.2 are examples that illustrate this confusion in engineering practice.

1.2 DIMENSIONS, UNITS, AND UNIT CONVERSIONS

Dimensions and units are not the same. They are two different categories. Dimensions are described as the categories of physical quantities that can be measured in a certain identification. For example, a dimension can be identified as a length or a mass. Units are described as a category of specified names that correlate to the measurement of the identified dimensions. For example, the length is a dimension, while a meter or a foot is a specific name (unit) of the measurement that describes the identified dimension of the length. Another example is time. Time is a dimension, while a second or a minute is a specific name (unit) of measurement that describes the identified dimension of the time. Hence, the unit is a standard of measurement chosen to quantify a corresponding identified dimension. Figure 1.1 shows the length and the mass in the category of dimensions. In the figure, the width and the height are the quantities that correspond to the dimensions of the length. The mass is the quantity that corresponds with the dimension within the quantity of volume. Figure 1.2 shows the units correlated to the standards of measurements of length and mass expressed as millimeters (mm) and kilogram (kg), respectively.

In engineering drawings, in addition to the objective shapes and system configurations being designed correctly, the measurement of the dimensions, also called dimensioning, must be indicated completely. Figure 1.3 are illustrations of dimensioning an object and a hot water heating piping system.

In engineering practice, the principle of dimension homogeneity and unit consistency needs to be held. Keeping the units consistent in equation calculations, such as in addition, abstraction, multiplication, and division unit conversions, is essential and necessary. Unit conversions convert units from one measurement to another or one measured unit system to another system. For example, 1 meter may need to be converted to 1,000 millimeters or 1 foot to be converted to 30.48 centimeters if the application requires.

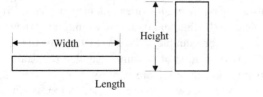

FIGURE 1.1 Dimensions.

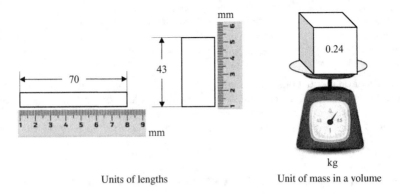

Units of lengths   Unit of mass in a volume

FIGURE 1.2   Units.

1.3   IMPORTANCE OF DIMENSIONS, UNITS, AND UNIT CONVERSIONS

Dimensions and units play an important role in engineering practice. Without proper dimensions and units to define quantifiable goods, there is no basis for commercialism or trade. Without showing dimensions and units in drawings, manufacturing will not be possible. Dimensions and units in calculations must be consistent, which is called the principle of dimension homogeneity and unit consistency. Different categories of dimensions and measurements of units cannot be manipulated in calculations. For example, the length of a block cannot be manipulated with the temperature of the block, and the mass of the block cannot be manipulated with the velocity of the block because they are different categories of dimensions. In calculations for the same category of dimensions, for example, in length calculations, meters cannot be manipulated with feet, and feet cannot be manipulated with inches directly because they are different units of measurement. In other words, like trying to add apples and oranges as shown in Figure 1.4, such a manipulation is not valid.

It is a common exercise to check the dimensional homogeneity and unit consistency in engineering practice. The dimensional analysis from the principle of dimensional homogeneity serves as a plausibility way for conducting proper calculations (see Section 2.3 in Chapter 2) and deriving the quantity relations (dimensionless numbers) for the complex physical phenomena in the absence of a rigorous theoretical description (see Chapter 5). Through unit conversions from the principle of unit consistency, those commensurable physical quantities that have the same dimension can be directly compared to each other and manipulated, even though they are originally expressed in different units, for example, meters to feet, grams to pounds, and hours to years. However, the manipulation of incommensurable physical quantities should be avoided no matter what units they are expressed in, such as meters to grams, seconds to grams, meters and seconds. For example, asking whether a gram is larger than an hour is meaningless.

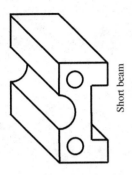

Short beam

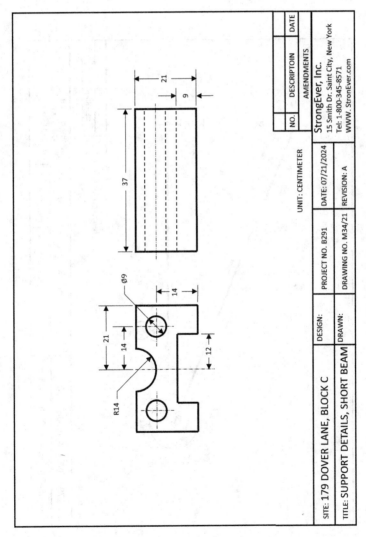

(a) Object in drawing

FIGURE 1.3 Dimensioning an object and a hot water piping system in drawings.

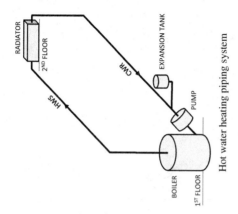

Hot water heating piping system

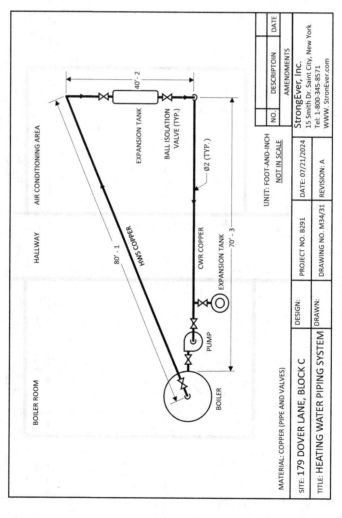

(b) Piping system in a drawing

FIGURE 1.3 *(Continued)*

FIGURE 1.4 Apples and oranges cannot be added.

In engineering practice, any calculations that are mathematically meaningful but physically meaningless, such as the numerical-value equation calculations, should be excluded. Briefly, the principle of the dimensional homogeneity and unit consistency helps

- express equations or physical expressions correctly,
- find relations between physical quantities to describe physical phenomena in the absence of a rigorous theoretical description, and
- convert units from one measurement to another or one unit system to another correctly (see Chapter 4).

Exercise 1.1

Kevin is a technician working in a furniture manufacturing factory. One day he is asked to manufacture a table with the dimensions of 6 long, 4 wide, and 3.5 height. Can Kevin make the table?

What is the table height? 3.5 feet or 3.5 meters?
What is the table area? 6 ft × 4 ft = 24 ft²
or 6 m × 4 m = 24 m²?

FIGURE *Exercise 1.1.*

SOLUTION

Kevin cannot manufacture a table with the dimensions of 6 long and 3.5 wide since he does not know the units. Will the table be 6 feet long, 4 feet wide, and 3.5 feet high or 6 meters long, 4 meters wide, and 3.5 meters high?

Exercise 1.2

Engineer Ian Taylor is conducting an energy consumption audit in a thermal power plant. He lets technician Smith Short from the plant's operation department go to check the temperature in a water tank. After checking, Smith comes back and reports to Ian that "the temperature is 45 degrees". Ian is not happy with the report. Why?

FIGURE *Exercise 1.2.*

SOLUTION

Because Smith does not tell what the temperature unit is being used, Ian does not know the temperature is 45 degrees Fahrenheit (°F) or Celsius (°C).

1.3.1 For Students Majoring in Engineering

In colleges and universities, it is necessary for students majoring in engineering to learn and understand dimensions, units, and unit conversions. In textbooks of fundamental engineering courses, such as statics, dynamics, thermodynamics, fluid mechanics, heat transfer, and others (the course list can be long), the beginning chapters typically need to introduce dimensions, units, and unit conversions. Students are educated to acknowledge that dimensions, units, and unit conversions play an imperative role in problem-solving and equation calculations. The textbooks commonly apply units in both the English system and the SI. Some of the example problems use units of the SI or the English system and some use mixtures of units or dual units. Consequently, unit conversions are frequently encountered to keep unit consistency. On class tests, quizzes, and exams, students are required to not only present calculation steps but also show the units in the steps. Students are educated to use unit consistency in calculations as a useful approach for checking whether a calculation result is correct and for finding where errors might occur in a calculation if the result is wrong.

The education for a solid knowledge of dimensions, units, and unit conversions is critical for today's students majoring in engineering. Students are expected to retain their understanding of dimensions, units, and unit conversions in both the English system and the SI and efficiently apply the knowledge of both systems in engineering practice after their graduation.

Dimensioning engineering drawings is an essential skill for students majoring in engineering. Without dimensions in engineering drawings, ideas and concepts in any design are impossible to understand. The drawings with symbols and data, such as dimensions, ensure that the communication between designers and producers is clear and precise. Students are educated to acknowledge that accurate dimensioning in engineering drawings helps the design be produced correctly and efficiently.

1.3.2 FOR ENGINEERS ENGAGING IN ENGINEERING PRACTICE

In today's commercial and engineering society, it is necessary more than ever for engineers engaging in engineering practice to thoroughly understand units and dimensions and apply skills for unit conversions. In engineering applications, correct and efficient communication of dimensions and units can avoid misunderstandings and misinterpretations in designs, productions, and R&D. Otherwise, it may cause time to be wasted, cost to be lost, and have significantly negative impact on the engineering activities.

The United States is a country that uses a dual-unit system, the English system and the SI, in engineering practice. Unfortunately, the English system is currently the prevailing unit system in U.S. companies. Unit conversions from one unit system to another are common and necessary in a company's routine work. There are many examples of engineers needing to convert the dimensions and units in the English system, in the SI, and between the two systems. For example, imported machines and equipment from other countries that use the units of the SI may need to convert technical data and specifications to the units of the English system. The conversion might be annoying, but it is a reality in engineering practice in the United States. Engineering literature, including not only papers, journals, magazines, and books but also reports, proposals, and presentations, commonly use dual units in the United States.

1.3.3 FOR THOSE PREPARING FOR THE FE/PE EXAMS

The Fundamentals of Engineering (FE) and Principles and Practice of Engineering (PE) exams are held by the National Council of Examiners for Engineering and Surveying (NCEES). On the exams, the knowledge of units, unit conventions, and engineering drawings is tested. The FE exam is the first exam in the process of becoming a licensed Professional Engineer (P.E.). The FE exam is designed for recent graduates and students who are close to finishing an undergraduate engineering degree from an EAC/ABET-accredited program. The exam is comprehensive and covers most subjects in an undergraduate engineering curriculum. The PE exam is the second exam in the process of becoming a P.E. The PE exam is designed for engineers who have gained a minimum of 4 years of post-college work experience in their chosen engineering discipline. The exam tests a minimum level of competency in a particular engineering discipline.

The FE and PE exams require students to have a sound understanding of the concepts of dimensions, units, and unit conversions in the English systems and the SI and efficiently use units and unit conversions for problem-solving. It should be acknowledged that the NCEES provides a reference handbook during the exams in which basic equations and specified design standards are shown. However, the equations and standards that appear in the handbook are more generic and do not necessarily differentiate between the unit application of the English system and the SI. Therefore, students must be careful, especially when applying units in the English system, to make sure the necessary conversion factors are being used for unit consistency, for example, the need to use the g_c conversion factor in calculations with the unit of force or any quantities related to force, such as torque, pressure, kinetic energy, potential energy, and others. Generally, the FE exam focuses on theory, whereas the PE exam is more practice-based. It has to be remembered that both the FE and PE exams use units of the English system and the SI. Knowledge of the drawings and dimensioning is tested in the FE and PE exams. In the FE exam of the mechanical discipline, the knowledge is tested in Engineering Drawing Interpretations and Geometric Dimensioning and Tolerancing (GD&T) under Topic 14: Mechanical Design and Analysis (see A.8a in the Appendix). In the PE mechanical exam for the thermal and fluid systems discipline, students' knowledge of units, conversions, and technical drawings is tested in A. Basic Engineering Practice under the Topic I. Principle (see A.8b in the Appendix).

Exercise 1.3 (FE style)

Find all correct area calculations expressed if the unit conversion is not conducted.

(A) 30 × 60
(B) 30 in. × 600 mm
(C) 10 ft × 15 ft
(D) 3 m × 6 m

SOLUTION

The correct answers are **(C)** and **(D)**. (A) is a numerical-value expression. In (B), the units are not consistent if a unit conversion is not conducted.

2 Dimensions

There are two kinds of dimensions of quantities. One is the fundamental dimensions, also known as the primary dimensions. Another is the derived dimensions, also known as the secondary dimensions. The derived dimensions can be obtained from the fundamental dimensions and expressed in the combination of the fundamental dimensions.

2.1 FUNDAMENTAL DIMENSIONS

The dimensions of quantities that are independent of other quantities are the fundamental dimensions (primary dimensions). There are seven types of fundamental dimensions based on the measurement types of the quantities: mass, length, time, temperature, electric current, luminous intensity, and amount of substance.

Instead of using the names of quantities to represent the dimensions, dimension symbols are commonly used. Table 2.1 presents the fundamental dimensions and their symbols. A particular engineering application may not need to use all of them. For example, in thermal fluids applications, only the first four in the table are required:

- Mass, $\{m\}$
- Length, $\{l\}$
- Time, $\{t\}$
- Temperature, $\{T\}$

The format of a quantity symbol with curly brackets used in the book means that the expression is the dimension symbol of the quantity. For example, the formats of

$$\{m\} = M$$

and

$$\{l\} = L$$

are used to express the dimension symbols. Similarly, the dimension of mass m is M and length l is L. If thermodynamic temperature is not involved in the application, capitalized T is used for the dimension symbol of time, such as in statics and dynamics. If both temperature and time appear in the application, t or the Greek letter θ is used for the dimension symbols of time and T for the dimension symbol of temperature, respectively.

DOI: 10.1201/9781003508977-2

Attention must be paid to distinguish the symbols of m (roman), m (italic), and M (capitalized italic). The symbol m is the unit (meter) in the SI. The symbol m is the quantity of mass. The symbol M is the dimension of mass. The SI proposes using "roman" for the dimension symbol. This book uses "*italic*" to draw more attention to it.

Regardless of the systems of dimensions and units chosen for measurement, the variable in the application that represents the same physical quantity should be the same symbol for the fundamental dimensions. For example, the dimension of mass in both the SI and English systems is the same symbol M. Absolute temperature and atmospheric temperature have the same dimension symbol T. While velocity is a derived dimension, which can be expressed in different units depending on which system is used, the SI or the English system, such as miles per hour (mil/h) or meters per second (m/s). However, the dimensions of both must be expressed in the same symbols with the fundamental dimensions, that is, L/t, length per unit time. The fundamental dimensions shown in Table 2.1 cannot be further broken down into more basic dimensions. The concept of fundamental dimensions is particularly useful in dimensional analysis.

Depending on which dimensions are chosen, there are three dimension systems in engineering applications:

- The $ML\theta T$ system
- The $FL\theta T$ system
- The $FML\theta TQ$ system

In the systems, the symbol θ denotes time, and the symbol T denotes temperature, respectively.

TABLE 2.1
Fundamental Dimensions and Symbols

Fundamental Dimensions		
Quantity		**Dimension**
Name	**Quantity Symbol**	**Symbol**
Mass	m	M
Length	l	L
Time	t	t (θ or T)
Thermodynamic temperature	T	T
Electric current	I	I
Luminous intensity	I_v	C
Amount of substance	n	N

Note: If both temperature and time are presented in the application, the dimension symbol of temperature is T, and time can be either t or θ.

MLθT system

The $ML\theta T$ system is the most popularly used system in engineering practice. It is also a default system in dimensional analysis. The fundamental dimensions defined in the $ML\theta T$ system are mass M, length L, time θ, and temperature T. Other dimensions of the physical quantities can be derived from these fundamental dimensions. For example, the dimension of work is ML^2/θ^2, which is a combination of the fundamental dimensions of mass M, length L, and time θ.

FLθT system

The $FL\theta T$ system is almost the same as the $ML\theta T$ system. The difference between the $FL\theta T$ system and the $ML\theta T$ system is that the fundamental dimension of mass M in the $ML\theta T$ system is replaced by the dimension of force F. Therefore, the dimensions defined in the $FL\theta T$ system are force F, length L, time T, and temperature θ. For example, work has a dimension of FL expressed by the dimensions of force F and length L.

FMLθTQ system

In certain applications, a particular dimension system may be more convenient to deal with the thermodynamic quantities, such as heat. Therefore, the $FML\theta TQ$ system is developed. The system adds two additional dimensions, force F and heat Q, based on the $ML\theta T$ system. Consequently, work has the dimension of FL which is the same as that in the $FL\theta T$ system. The dimensions of thermodynamics quantities can be expressed in simpler formats. For example, the dimensions of heat and heat flowrate are expressed as a Q directly and $Q\theta^{-1}$, respectively, instead of $ML^2\theta^{-2}$ and $ML^2\theta^{-3}$ in the $ML\theta T$ system.

Table 2.2 shows the dimensions of commonly used quantities in the three systems in engineering applications. For the convenience of reading and comparison, some derived dimensions (see Section 2.2) are also presented. Whether the $ML\theta T$, $FL\theta T$, or $FML\theta TQ$ system is being used in engineering practice depends on the application circumstances. For example, the $FL\theta T$ system may be convenient to use when force dominates in applications, such as in statics and dynamics. The $FML\theta TQ$ system may be more convenient to use when heat is involved, such as in applications of thermodynamics and heat transfer. If the $ML\theta T$ system is used in the English systems, the unit conversion constants, such as the gravitational constant gc and mechanical equivalent of heat J, are almost certainly involved to keep unit consistency among the quantities.

Exercise 2.1

Verify the dimension of gas constant R shown in Table 2.2.

SOLUTION

The gas constant R is defined as

$$R = \frac{R_u}{M}$$

TABLE 2.2

Dimensions of Some Commonly Used Quantities in the Three Systems

Quantity		Dimension		
Name	Symbol	$ML\theta T$	$FL\theta T$	$FML\theta TQ$
Mass	m	M	$F\theta^2 L^{-1}$	M
Length	l	L	L	L
Time	t	θ	Θ	θ
Temperature	T	T	T	T
Force	F	$ML\theta^{-2}$	F	F
Heat	Q	$ML^2\theta^{-2}$	FL	Q
Acceleration	a	$L\theta^{-2}$	$L\theta^{-2}$	$L\theta^{-2}$
Angular velocity	ω	θ^{-1}	θ^{-1}	θ^{-1}
Area	A	L^2	L^2	L^2
Coefficient of thermal expansion	β	T^{-1}	T^{-1}	T^{-1}
Density	ρ	ML^{-3}	$FL^{-4}\theta^2$	$FL^{-4}\theta^2$
Energy	E	$ML^2\theta^{-2}$	FL	FL
Frequency	f	θ^{-1}	θ^{-1}	θ^{-1}
Friction	f	$ML\theta^{-2}$	F	F
Gas constant	R	$L^2\theta^{-2}T^{-1}$	$L^2\theta^{-2}T^{-1}$	$QM^{-1}T^{-1}$
Gravitational acceleration	g	$L\theta^{-2}$	$L\theta^{-2}$	$L\theta^{-2}$
Gravitational constant	g_c	–	–	$ML\theta^{-2}F^{-1}$
Heat flow rate	$\dot{Q}$	$ML^2\theta^{-3}$	$FL\theta^{-1}$	$Q\theta^{-1}$
Heat transfer coefficient of convection	h	$M\theta^{-3}T^{-1}$	$F\theta^{-1}L^{-1}T^{-1}$	$Q\theta^{-1}L^{-2}T^{-1}$
Height	z	L	L	L
Kinematic viscosity	v	$L^2\theta^{-1}$	$L^2\theta^{-1}$	$L^2\theta^{-1}$
Mass flow rate	$\dot{m}$	$M\theta^{-1}$	$F\theta L^{-1}$	$M\theta^{-1}$
Mechanical equivalent of heat	J	–	–	FLQ^{-1}
Molar mass	M	$M\,N^{-1}$	$F\theta^2 L^{-1}N^{-1}$	MN^{-1}
Moment	M	$ML^2\theta^{-2}$	FL	FL
Overall Heat transfer coefficient	U	$M\theta^{-3}T^{-1}$	$F\theta^{-1}L^{-1}T^{-1}$	$Q\theta^{-1}L^{-2}T^{-1}$
Power	P	$ML^2\theta^{-3}$	$FL\theta^{-1}$	$FL\theta^{-1}$
Pressure	p	$ML^{-1}\theta^{-2}$	FL^{-2}	FL^{-2}
Specific heat at constant pressure	c_p	$L^2\theta^{-2}T^{-1}$	$L^2\theta^{-2}T^{-1}$	$QM^{-1}T^{-1}$
Specific heat at constant volume	c_v	$L^2\theta^{-2}T^{-1}$	$L^2\theta^{-2}T^{-1}$	$QM^{-1}T^{-1}$
Specific volume	v	L^3M^{-1}	$L^4F^{-1}\theta^{-2}$	L^3M^{-1}
Specific weight	γ	$ML^{-2}\theta^{-2}$	FL^{-3}	FL^{-3}
Surface tension	σ	$M\theta^{-2}$	FL^{-1}	FL^{-1}
Thermal conductivity	k	$ML\theta^{-3}T^{-1}$	$F\theta^{-1}T^{-1}$	$QL^{-1}\theta^{-1}T^{-1}$

TABLE 2.2 (Continued)
Dimensions of Some Commonly Used Quantities in the Three Systems

Quantity		Dimension		
Name	Symbol	$ML\theta T$	$FL\theta T$	$FML\theta TQ$
Thermal diffusivity	α	$L^2\theta^{-1}$	$L^2\theta^{-1}$	$L^2\theta^{-1}$
Torque	T	$ML^2\theta^{-2}$	FL	FL
Velocity	V	$L\theta^{-1}$	$L\theta^{-1}$	$L\theta^{-1}$
Volume	V	L^3	L^3	L^3
Volumetric flow rate	$\dot{V}$	$L^3\theta^{-1}$	$L^3\theta^{-1}$	$L^3\theta^{-1}$
Viscosity	μ	$ML^{-1}\theta^{-1}$	$FL^{-2}\theta$	$FL^{-2}\theta$
Universal gas constant	R_u	$ML^2\theta^{-2}T^{-1}N^{-1}$	$FLT^{-1}N^{-1}$	$QT^{-1}N^{-1}$
Weight	W	$ML\theta^{-2}$	F	F
Work	W	$ML^2\theta^{-2}$	FL	FL

Note: In the table, the symbol θ is the dimension of time, and T is the dimension of temperature.

where R_u is the universal gas constant and M is the gas molar mass, respectively.
Referring to Table 2.2, quantities of R_u and M in the $ML\theta T$ system have the dimensions

$$\{R_u\} = ML^2\theta^{-2}T^{-1}N^{-1}$$

and

$$\{M\} = MN^{-1}$$

Therefore, the dimension of R in the $ML\theta T$ system is determined to be

$$\{R\} = \frac{ML^2\theta^{-2}T^{-1}N^{-1}}{MN^{-1}} = L^2\theta^{-2}T^{-1}$$

While the quantities of R_u and M in the $FL\theta T$ system have the dimensions

$$\{R_u\} = FLT^{-1}N^{-1}$$

and

$$\{M\} = F\theta^2L^{-1}N^{-1}$$

the dimension of R in the $FL\theta T$ system is determined to be

$$\{R\} = \frac{FLT^{-1}N^{-1}}{F\theta^2L^{-1}N^{-1}} = L^2\theta^{-2}T^{-1}$$

In the $FML\theta TQ$ system, the quantities of R_u and M have the dimensions

$$\{R_u\} = QT^{-1}N^{-1}$$

and

$$\{M\} = MN^{-1}$$

Therefore, the dimension of R in the $FML\theta TQ$ system is determined to be

$$\{R\} = \frac{QT^{-1}N^{-1}}{MN^{-1}} = QM^{-1}T^{-1}$$

Exercise 2.2 (FE style)

In the $ML\theta T$ system, the dimension of thermal conductivity k (W/m.K) is

(A) $ML^{-1}\theta^{-1}T^{-2}$.
(B) $ML\theta T^{-1}$.
(C) $ML\theta^2 T^{-1}$.
(D) $ML\theta^{-3}T^{-1}$.

SOLUTION

The correct answer is **(D)**. Referring to Table 2.2 or knowing that the dimensions of thermal conductivity k (W/m·K) are $ML^2\theta^{-3}$, L, and θ for power (W), meter (m), and absolute temperature (K), respectively, therefore,

$$\{k\} = \frac{ML^2\theta^{-3}}{(L)(T)} = ML\theta^{-3}T^{-1}$$

Exercise 2.3 (FE style)

Find all of the following fundamental dimensions:

(A) Mass
(B) Time
(C) Force
(D) Amount of substance

SOLUTION

The correct answers are **(A)**, **(B)**, and **(D)**. Force is a derived dimension.

2.2   DERIVED DIMENSIONS

Except for the fundamental (primary) dimensions shown in Table 2.1, all other dimensions are the derived (secondary) dimensions. The derived dimensions are derived from the fundamental dimensions and are expressed in a combination of the fundamental dimensions. For example, the dimension of force in the $ML\theta T$ system is expressed by the fundamental dimensions of mass M, length L, and time θ, that is, $ML\theta^{-2}$. Some derived dimensions commonly used in the $ML\theta T$ system without temperature presented are shown in Table 2.3.

TABLE 2.3
Some Derived Dimensions in the $ML\theta T$ System

Quantity		Dimension
Name	**Symbol**	**$ML\theta T$**
Acceleration	a	LT^{-2}
Density	ρ	ML^{-3}
Energy, Heat, Work	E, Q, W	ML^2T^{-2}
Force	F	MLT^{-2}
Friction	f	MLT^{-2}
Gravitational acceleration	g	LT^{-2}
Kinematic viscosity	ν	L^2T^{-1}
Moment, Torque	M	ML^2T^{-2}
Power	P	ML^2T^{-3}
Pressure	p	$ML^{-1}T^{-2}$
Specific weight	γ	$ML^{-2}T^{-2}$
Time	t	T
Velocity	V	LT^{-1}
Viscosity, Absolute viscosity	μ	$ML^{-1}T^{-1}$

Note: In the table, the symbol T is the dimension of time.

Example 2.4

A vehicle having a mass m moves along a horizontal road at a constant velocity V. Determine the dimension of the vehicle's linear momentum.

SOLUTION

The vehicle's linear momentum $\vec{G}$ is expressed as

$$\vec{G} = m\vec{V}$$

In the equation, m is mass and $\vec{V}$ is velocity. The arrow symbol "→" above the variable means that the variable is a vector. Their dimensions are

$$\{m\} = M$$

and

$$\{\vec{V}\} = LT^{-1}$$

respectively. The dimension of the linear momentum $\vec{G}$ of the vehicle, therefore, is

$$\{\vec{G}\} = MLT^{-1}$$

Exercise 2.5

An airplane having a mass m (kg) is cruising at a speed V (m/s) at an altitude h (m). Determine the dimension of the total energy of the airplane by using (a) the $ML\theta T$ system, (b) the $FL\theta T$ system, and (c) the $FML\theta TQ$ system, respectively.

SOLUTION

The total energy of the cruising airplane is calculated by the equation

$$E = \frac{1}{2}mV^2 + mgh$$

(a) In the $ML\theta T$ system, the dimensions of mass, speed, gravitational acceleration, and altitude are

$$\{m\} = M$$
$$\{V\} = LT^{-1}$$
$$\{g\} = LT^{-2}$$
$$\{h\} = L.$$

Therefore, the dimension of the total energy of the cruising airplane is

$$\{E\} = \frac{1}{2}(M)(LT^{-1})^2 + (M)(LT^{-2})(L)$$
$$= \frac{1}{2}(M)(LT^{-1})^2 + (M)(LT^{-2})(L)$$
$$= \frac{1}{2}ML^2T^{-2} + ML^2T^{-2}$$
$$\{E\} = ML^2T^{-2}$$

(b) In the $FL\theta T$ system, the dimensions of mass, speed, gravitational acceleration, and altitude are

$$\{m\} = FT^2L^{-1}$$
$$\{V\} = LT^{-1}$$
$$\{g\} = LT^{-2}$$
$$\{h\} = L$$

Therefore, the dimension of the total energy of the cruising airplane is

$$\{E\} = \frac{1}{2}(FT^2L^{-1})(LT^{-1})^2 + (FT^2L^{-1})(LT^{-2})(L)$$

$$= \frac{1}{2}FL + FL$$

$$\{E\} = \textbf{FL}$$

(c) In the $FML\theta TQ$ system, the dimensions of mass, speed, gravitational acceleration, and altitude are

$$\{m\} = M$$

$$\{V\} = LT^{-1}$$

$$\{g\} = LT^{-2}$$

$$\{h\} = L.$$

Therefore, the dimension of the total energy of the cruising airplane is

$$\{E\} = \frac{1}{2}(M)(LT^{-1})^2 + (M)(LT^{-2})(L)$$

$$= \frac{1}{2}ML^2T^{-2} + ML^2T^{-2}$$

Since

$$\{F\} = MLT^{-2}$$

$$\{E\} = \textbf{FL}$$

Exercise 2.6 (FE style)

Find all derived dimensions for the following and show their dimensions in fundamental dimensions.

(A) Temperature T
(B) Velocity V
(C) Density ρ
(D) Work W

SOLUTION

The correct answers are **(B)**, **(C)**, and **(D)**. Temperature is a fundamental dimension. From Table 2.3, the dimensions of velocity V, density ρ, and Work $\underline{W}$ are LT^{-1},
ML^{-3}, and ML^2T^{-2}, respectively.

2.3   THE PRINCIPLE OF DIMENSIONAL HOMOGENEITY

The principle of dimensional homogeneity is a basic axiom in dimensional analysis. The principle states:

> *Every additive term in a physical expression and an equation in calculation must have the same dimension.*

In other words, any quantity in physical expressions and equations containing different dimensions cannot be added and subtracted. The dimensions of the left side and right side in an equation must have identical dimensions. For example, in the expression $V_1^2 = V_2^2 + 2as$ for calculating a particle rectilinear motion, the dimensions of V_1^2 on the left side and V_2^2 and $2as$ on the right side are the same dimensions of L^2T^{-2}. This principle of dimensional homogeneity also holds for the units, which is well known as the principle of unit consistency.

Exercise 2.7

Verify that the ideal gas equation of state

$$Pv = RT$$

is an equation of dimensional homogeneity. In the equation, P is the absolute pressure, v is the specific volume, R is the gas constant, and T is the absolute temperature.

SOLUTION

Referring to Table 2.2, the dimensions of the pressure P, the specific volume v, the gas constant g, and the temperature T are, respectively,

$$\{P\} = ML^{-1}\theta^{-2},$$
$$\{v\} = L^3 M^{-1},$$
$$\{R\} = L^2 \theta^{-2} T^{-1},$$
$$\{T\} = T.$$

The dimensions on both sides of the equation, therefore, are

$$(ML^{-1}\theta^{-2})(L^3 M^{-1}) = (L^2 \theta^{-2} T^{-1})(T)$$

that is,

$$L^2 \theta^{-2} = L^2 \theta^{-2}$$

As a result, the ideal gas equation of state is an equation of dimensional homogeneity.

Exercise 2.8

An algebra mathematical equation is described as

$$Z = x + y \qquad\qquad (E2.1a)$$

Knowing that both x and y are the quantities in a linear movement, x is the displacement quantity of a uniformly accelerated body starting from rest, and y is the quantity of velocity over time. Accordingly,

$$x = \frac{1}{2}at^2 \qquad\qquad (E2.1b)$$

$$y = at \qquad\qquad (E2.1c)$$

where a is the uniform acceleration and t is the time variable. Replacing x and y by Equations (E2.1b) and (E2.1c), Equation (E2.1a), therefore, can be expressed as

$$Z = \frac{1}{2}at^2 + at \qquad\qquad (E2.1d)$$

SOLUTION

Equation (E2.1d) mathematically is correct. But the equation is physically meaningless since the equation sums incompatible quantities of displacement and velocity. After showing the dimensions of x and y,

$$\{x\} = \frac{1}{2}\left(LT^{-2}\right)\left(T^2\right) = L$$
$$\{y\} = \left(LT^{-2}\right)(T) = LT^{-1}$$

It can be seen the dimension of Z cannot be determined. Equation (E2.1d) obviously violates the principle of dimensional homogeneity. Therefore, Z cannot be a summation of x (displacement) and y (velocity). Equation (E2.1d) is physically wrong.

Therefore, the principle of dimensional homogeneity is useful when checking the algebra of a problem solution. Namely, the dimensional inconsistency of an equation is a sure sign of an algebraic error in the calculation.

Exercise 2.9 (FE style)

Find all expressions shown that <u>do not comply</u> with the principle of dimensional homogeneity.

(A) $15 + 26$

(B) $P + \frac{1}{2}\rho V^2 + \rho gz$

(C) $kA\dfrac{T_1-T_2}{l}+hA(T_s-T_\infty)$

(D) $y_0+V_0t-\dfrac{1}{2}gt^2$

SOLUTION

The correct answer is **(A)** since the expression is a numerical-value equation. (B), (C), and (D) are in dimensional homogeneity.

(B)   Referring to Table 2.3, since the expression has no temperature involved, the dimensions of quantities are

$$\{P\}=ML^{-1}T^{-2}$$

$$\left\{\frac{1}{2}\rho V^2\right\}=\left(L^{-3}M\right)\left(LT^{-1}\right)^2=ML^{-1}T^{-2}$$

$$\{\rho gz\}=\left(L^{-3}M\right)\left(LT^{-2}\right)(L)=ML^{-1}T^{-2}.$$

The dimensions on the left side and the right side of the expression are identical. Therefore, the expression is in dimensional homogeneity. The expression actually is the Bernoulli equation in fluid mechanics.

(C)   Referring to Table 2.2, the dimensions of quantities are

$$\left\{kA\frac{T_1-T_2}{l}\right\}=(ML\theta^{-2}T^{-1})\left(L^2\right)\left(TL^{-1}\right)=ML^2\theta^{-2}$$

$$\{hA(T_s-T_\infty)\}=(M\theta^{-2}T^{-1})\left(L^2\right)(T)=ML^2\theta^{-2}.$$

The dimensions on the left side and the right side of the expression are identical. Therefore, the expression is in dimensional homogeneity. The expression actually is the heat transfer rate $\dot{Q}$ of mixing conduction and convection in heat transfer.

(D)   Referring to Table 2.3, since the expression has no temperature involved, the dimensions of quantities are

$$\{y_0\}=L$$

$$\{V_0t\}=\left(LT^{-1}\right)(T)=L$$

$$\left\{\frac{1}{2}gt^2\right\}=\frac{1}{2}\left(LT^{-2}\right)\left(\{T^2\}\right)=L.$$

The dimensions on the left side and the right side of the expression are identical. Therefore, the expression is in dimensional homogeneity. The expression actually is the equation of projectile motion in dynamics.

3 Units

As described in Section 1.1 (see Chapter 1), any physical quantity can be expressed as a combination of a numerical value and a unit of measurement,

$$Q = \mathrm{N}q$$

Quantity   Numerical value   Unit

Without the unit, the quantity is meaningless in engineering applications. Units also have two types that correspond to the fundamental and derived dimensions. One is fundamental units, also called primary units, and another is derived units, also called secondary units. Over the years, a number of unit systems have been developed. In the United States, two unit systems are commonly used. They are the English system, known as the U.S. Customary Units (USCU), and the International System (SI) of Units known as the metric units. English system units are widely used on consumer products and in manufacturing. SI units are the standard in science and medicine, as well as many sectors of industry and government, including the military. In both systems, there are further distinguished subsystems. In the English system are

- the English Engineering System,
- the English Gravitational System, and
- the Absolute English System.

In the SI are

- the CGS system and
- the MKS system.

3.1 FUNDAMENTAL AND DERIVED UNITS

The units used for measurements of the fundamental dimensions are called the fundamental (primary) units, and the units used for measurements of the derived dimensions are called the derived (secondary) units. The fundamental and derived units are further distinguished into the units of the English system and the SI. Some commonly used fundamental units and derived units in both systems in engineering practice are shown in Table 3.1 and Table 3.2, respectively.

DOI: 10.1201/9781003508977-3

TABLE 3.1
Commonly Used Fundamental Units in the English System and the SI

Quantity	English Unit		SI Unit	
	Name	Symbol	Name	Symbol
Mass m	pound-mass	lbm	kilogram	kg
Length l	foot	ft	meter	m
Time t	second	s	second	s
Temperature T	Rankine	R	kelvin	K

TABLE 3.2
Commonly Used Derived Units in the English System and the SI

Quantity	English Unit		SI Unit	
	Name	Symbol	Name	Unit
Acceleration a	foot per square second	(ft/s^2)	meter per second squared	(m/s^2)
Density ρ	pound-mass per cubic foot	(lbm/ft^3)	kilogram per cubic meter	(kg/m^3)
Force F	pound-force	lbf	newton	N
Energy E Heat Q Work W	British thermal unit	Btu	joule	J
Power P	horsepower	hp	watt	W
Pressure P	pound-force per square inch	psi	pascal	Pa
Velocity V	feet per second	(ft/s)	meter per second	(m/s)
Volumetric flow rate $\dot{V}$	cubic foot per second	(ft^3/s)	cubic meter per second	(m^3/s)

Note: The unit symbol in parentheses is a combination of the fundamental units.

There are several general rules regarding the formats of unit names and symbols:

- The absolute temperature units, R and K, do not need to show the degree symbol, "°", with them.
- The unit names are in the lowercase even though they are derived from people's names. However, the unit symbol should be capitalized if it is derived from a person's names. For example, in SI units, the name of absolute temperature named after Lord Kelvin (1824–1907) is kelvin, and the symbol is K.

- The full name of a unit may be pluralized, but its symbol is never pluralized. For example, mass in SI units can be expressed as 10 kilograms but not as 10 kgs.
- No periods are required in the unit symbols in a word line unless they appear at the end of a sentence. For example, in the SI, the phrase "10 m long" is correct, but "10 m. long" is not.

3.2 THE ENGLISH SYSTEM

The English system used in the United States, known as the USCU system, developed in 1790. The system is primarily based on historical and customary units that have evolved over time. In 1832, the system was standardized and adopted in the United States. The English system relies on various non-decimal conversions between the units for measuring different quantities. For example, inches, feet, yards, and miles are the commonly used units for measuring length: 12 inches in a foot, 3 feet in a yard, and 5,280 feet in a mile. The units of measurement constituting from different amounts make unit conversions hard.

There are three subsystems in the English system. They are the English Engineering System, the English Gravitational System, and the Absolute English System. Nevertheless, the root of defining the units in the three systems is from Newton's second law

$$F = ma \tag{3.1}$$

In the equation, F is the quantity of force acting on an object, m is the quantity of mass the object has, and a is the quantity of acceleration of the object as a result of the acting force.

3.2.1 THE ENGLISH ENGINEERING SYSTEM

In the English Engineering System, the units defined from Newton's second law are independent. The unit of the force F is lbf, the mass m is lbm, and the acceleration a is ft/s². According to the principle of unit consistency, the units of terms on the left side and right side in Equation (3.1) must be the same. Unfortunately, they are not. In order to make unit consistent in the equation, the unit conversion constant gc in the unit of lbm·ft/lbf·s² called the gravitational constant is used in the equation. Consequently, the units of Newton's second law in the English Engineering System becomes consistent, that is,

$$F \, \mathrm{lbf} = \frac{(m \, \mathrm{lbm})\left(a \dfrac{\mathrm{ft}}{\mathrm{s}^2}\right)}{g_c \dfrac{\mathrm{lbm \cdot ft}}{\mathrm{lbf \cdot s}^2}} \tag{3.2}$$

The numerical value of g_c is 32.174.

Exercise 3.1

A force acts on a block. As a result, the block with a mass of 10 lbm moves at an acceleration of 20 ft/s² linearly. Determine the force acting on the block in lbf.

SOLUTION

If simply using Equation (3.1),

$$F = ma$$

and substituting the values of m and a in the equation, the unit conversion factor of g_c

$$F = (10\,1\text{bm})\left(20\,\frac{\text{ft}}{\text{s}^2}\right) \neq 200\,\text{lbf}$$

Obviously, the answer is not correct. This is a typical example of a mistake using Newton's second law without considering unit consistency. To get the correct answer by using Equation (3.2), the gravitational constant g_c should be used:

$$F\,1\text{bf} = \frac{(m\,\text{lbm})\left(a\,\dfrac{\text{ft}}{\text{s}^2}\right)}{g_c\,\dfrac{\text{lbm·ft}}{\text{lbf·s}^2}}$$

Substituting

$$m = 10\,\text{lbm}$$
$$a = 20\,\frac{\text{ft}}{\text{s}^2}$$

into the equation, then,

$$F = \frac{(10\,\text{lbm})\left(20\,\dfrac{\text{ft}}{\text{s}^2}\right)}{32.174\,\dfrac{\text{lbm·ft}}{\text{lbf·s}^2}} = \textbf{6.216 lbf}$$

Exercise 3.2

Use the equation of weight

$$W = mg$$

to determine the weight of an object with a mass of 20 lbm on Rarth. In the equation, g is the constant of gravitational acceleration. The quantity g is 32.174 ft/s².

SOLUTION

Referring to Equation (3.2), the weight of the object is determined to be

$$W \text{ lbf} = \frac{(m \text{ lbm})\left(g\,\frac{\text{ft}}{\text{s}^2}\right)}{g_c\,\frac{\text{lbm·ft}}{\text{lbf·s}^2}}$$

$$= \frac{(20 \text{ lbm})\left(32.174\,\frac{\text{ft}}{\text{s}^2}\right)}{32.174\,\frac{\text{lbm·ft}}{\text{lbf·s}^2}} = \textbf{20 lbf}$$

Exercise 3.2 presents an important fact, in the English engineering system, the numerical value of the mass in the unit lbm of an object is identical to that of its weight in the unit lbf on earth since the numerical values of g and g_c are the same. But attention must be paid to g and g_c having different physical meanings and units. The unit of the gravitational acceleration g is ft/s², and the unit of the gravitational constant g_c is lbm·ft/lbf·s².

Some quantities that the gravitational constant g_c lbm·ft/lbf·s² have to be applied in order to keep unit consistency in the English Engineering System:

- Kinetic energy

$$KE = \frac{\dot{m}V^2}{2g_c}\,\frac{\text{ft·lbf}}{\text{s}} \tag{3.3a}$$

$$ke = \frac{V^2}{2g_c}\,\frac{\text{ft·lbf}}{\text{lbm}} \tag{3.3b}$$

or

$$KE = \frac{\dot{m}V^2}{2g_cJ}\,\frac{\text{Btu}}{\text{s}} \tag{3.3c}$$

$$ke = \frac{V^2}{2g_cJ}\,\frac{\text{Btu}}{\text{lbm}} \tag{3.3d}$$

where J is a constant of the mechanical equivalent of heat, $J = 778$ ft·lbf/Btu.

- Potential energy

$$PE = \frac{\dot{m}gz}{g_c}\,\frac{\text{ft·lbf}}{\text{s}} \tag{3.4a}$$

$$pe = \frac{gz}{g_c}\,\frac{\text{ft·lbf}}{\text{lbm}} \tag{3.4b}$$

or

$$PE = \frac{\dot{m}gz}{g_c J}\frac{\text{Btu}}{\text{s}}$$

(3.4c)

$$pe = \frac{gz}{g_c J}\frac{\text{Btu}}{\text{lbm}}$$

(3.4d)

where z is the elevation unit in ft.

• Static pressure

$$P = \frac{g\rho h}{g_c}\frac{\text{lbf}}{\text{ft}^2}$$

(3.5)

where ρ is the density of the fluid in the unit lbm/ft³ and h is the height of the fluid in the unit ft.

• Weight

$$W = \frac{mg}{g_c}\text{1bf}$$

(3.6a)

or

$$\gamma = \frac{\rho g}{g_c}\frac{\text{lbf}}{\text{ft}^3}$$

(3.6b)

where γ is a specific weight defined as the weight per unit volume.

Exercise 3.3

A plate of 3 ft long and 2 ft wide is steadily floating in water horizontally as shown in Figure *Exercise 3.3*. Determine the force acting on the top of the plate.

SOLUTION

Knowing water density ρ = 62.4 lbm/ft³ and referring to Equation (3.5), the pressure acting on the top of the plate is

$$P = \frac{g\rho h}{g_c}$$

$$= \frac{\left(32.174\,\frac{\text{ft}}{\text{s}^2}\right)\left(62.4\,\frac{\text{lbm}}{\text{ft}^3}\right)(12\,\text{ft})}{32.174\,\frac{\text{lbm·ft}}{\text{lbf·s}^2}}$$

$$= 748.8\,\frac{\text{lbf}}{\text{ft}^2}$$

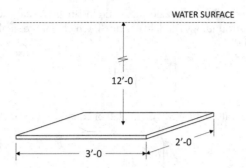

WATER SURFACE

12'-0

3'-0

2'-0

FIGURE *Exercise.3.3.*

The force is determined by

$$F = PA$$

where A is the surface area of the plate,

$$A = 3 \text{ ft} \times 2 \text{ ft} = 6 \text{ ft}^2$$

The force acting on the top of the plate floating in the water, therefore, is

$$F = (748.8 \frac{\text{lbf}}{\text{ft}^2})(6 \text{ ft}^2)$$
$$= 4,492.8 \text{ lbf}$$

Exercise 3.4

In a steam turbine, nozzles (stator) are used to convert the thermal energy of the steam flow to the kinetic energy of the steam flow acting on blades (rotor). The process of energy change per unit mass of the steam in a nozzle can be expressed by the energy balance equation,

$$q - w = \left(h + \frac{V^2}{2}\right)_2 - \left(h + \frac{V^2}{2}\right)_1 \qquad \text{(E3.4a)}$$

In the English system,

q is the heat exchange between the steam in the nozzle and surrounding the outside of the nozzle (Btu/lbm);

w is the specific work done by the nozzle (Btu/s);

h_1 and h_2 are the specific enthalpy of steam into and out of the nozzle, respectively (Btu/lbm); and

V_1 and V_2 are the velocity of the steam inlet to and outlet of the nozzle, respectively (ft/s).

Since no work is done by the nozzle and assuming that the heat exchange between the nozzle and surroundings is relatively small, which can be neglected, Equation (E3.4a) is simplified as

$$\left(h+\frac{V^2}{2}\right)_2 = \left(h+\frac{V^2}{2}\right)_1 \qquad \text{(E3.4b)}$$

It can be seen the units of the terms on the left of Equation (E3.4b) are not consistent,

$$\left(h+\frac{V^2}{2}\right)_2$$

$$Btu/1bm \neq \left(\frac{ft}{s}\right)^2$$

Therefore, the unit conversion constant g_c (lbm·ft/lbf·s²) and J (ft·lbf/Btu) must be applied to the term $V^2/2$ in Equation (E3.4b). So, Equation (E3.4b) becomes

$$\left(h+\frac{V^2}{2g_cJ}\right)_2 = \left(h+\frac{V^2}{2g_cJ}\right)_1 \qquad \text{(E3.4c)}$$

In Equation (E3.4c), the units are consistent now in Btu/lbm. To further simplify the equation in a typical application of the nozzle, considering the velocity out of the nozzle is much higher than the velocity into the nozzle, that is,

$$V_2 >> V_1$$

V_1 can be neglected. Therefore, the velocity out of the nozzle is determined as

$$V_2^2 = 2g_cJ(h_1 - h_2)$$
$$V_2 = \sqrt{2g_cJ(h_1 - h_1)} \qquad \text{(E3.4d)}$$

Now, considering that a steam stream accelerates in the nozzle having the outlet diameter $d = 2$ in. from the inlet pressure $P_1 = 400$ psia and the temperature $T_1 = 800\ °F$ to the exit saturated pressure P_2 in an isentropic process as shown in Figure Exercise 3.4 and neglecting any energy losses, estimate (a) the pressure P_2 and temperature T_2 of the steam out of the nozzle, (b) the exit steam velocity V in ft/s, (c) the mass flow rate $\dot{m}$ at the exit of the nozzle in lbm/s, and (d) the kinetic energy rate produced by the steam in the nozzle in Btu/s.

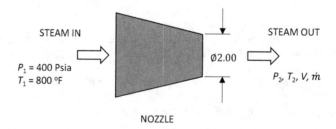

STEAM IN

$P_1 = 400$ Psia
$T_1 = 800\ °F$

Ø2.00

STEAM OUT

$P_2, T_2, V, \dot{m}$

NOZZLE

FIGURE   Exercise 3.4.

SOLUTION

Knowing the process from state 1 to state 2 at the saturated condition is an isentropic process, the process in the *T-s* diagram is illustrated as shown.

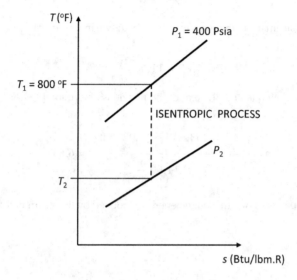

(a) The steam condition of $P_1 = 400$ Psia and $T_1 = 800$ °F tells that the steam is at the superheated state. P_2 is given at the saturated pressure. From the steam table, their enthalpies are found to be

$$h_1 = 1,417.00 \text{ Btu}/\text{lbm}$$

$$h_2 = 1167.76 \text{ Btu}/\text{lbm},$$

respectively. The pressure and temperature of the exit steam out of the nozzle are

$$P_2 = 36.06 \text{ Psia}$$

$$T_2 = 260.66°F$$

(b) Using Equation (E3.4d), the exit steam velocity out of the nozzle is

$$V_2 = \sqrt{2 g_c J (h_1 - h_2)}$$

$$= \sqrt{2\left(32.174 \frac{\text{lbm·ft}}{\text{lbf·s}^2}\right)\left(778 \frac{\text{ft·lbf}}{\text{Btu}}\right)(1,417.00 - 1,167.76)\frac{\text{Btu}}{\text{lbm}}}$$

$$= \sqrt{12,477,638.32 \frac{\text{ft}^2}{\text{s}^2}} = 3,532.37 \frac{\text{ft}}{\text{s}}$$

(c) The exit sectional area of the nozzle is

$$A = \frac{\pi d^2}{4} = \frac{3.14(2 \text{ in})^2}{4\left(144 \frac{\text{in}}{\text{ft}}\right)} = 0.0218 \text{ ft}^2$$

From the steam table, the specific volume of the steam at the exit $P_2 = 36.06$ Psia is

$$v_2 = 11.60 \frac{\text{ft}^3}{\text{lbm}}$$

The mass flow rate $\dot{m}$ at the exit of the nozzle is determined to be

$$\dot{m} = \frac{V_2 A}{v_2} = \frac{\left(3,532.37 \frac{\text{ft}}{\text{s}}\right)(0.0218 \text{ ft}^2)}{11.60 \frac{\text{ft}^3}{\text{lbm}}} = 6.63 \frac{\text{lbm}}{\text{s}}$$

Using Equation (3.3c), the kinetic energy produced by the steam in the nozzle is.

$$KE = \frac{\dot{m} V^2}{2 g_c J}$$

$$= \frac{\left(6.63 \frac{\text{lbm}}{\text{s}}\right)\left(3,532.37 \frac{\text{ft}}{\text{s}}\right)^2}{2\left(32.174 \frac{\text{lbm·ft}}{\text{lbf·s}^2}\right)\left(778 \frac{\text{ft·lbf}}{\text{Btu}}\right)}$$

$$= 1,652.46 \frac{\text{Btu}}{\text{s}}$$

Exercise 3.5 (FE style)

The work that moves an object having a mass of 10 lbm at an acceleration of 10 ft/s² ft for a distance of 10 ft will be

(A) 31.081 lbf·ft.
(B) 42.132 lbf·ft.
(C) 101.937 lbf·ft.
(D) 1,000 lbf·ft.

SOLUTION

The correct answer is (A). Work P is determined as

$$P = (F \text{ 1bf})(\text{l ft})$$

Using Equation (3.2),

$$F\ \text{lbf} = \frac{(m\ \text{lbm})\left(a\dfrac{\text{ft}}{\text{s}^2}\right)}{g_c\dfrac{\text{lbm·ft}}{\text{lbf·s}^2}}$$

therefore, the work is

$$P = \frac{(10\ \text{lbm})\left(10\dfrac{\text{ft}}{\text{s}^2}\right)}{32.174\dfrac{\text{lbm·ft}}{\text{lbf·s}^2}}(10\ \text{ft}) = \frac{1{,}000}{32.174}\text{lbf·ft}$$

$$= \mathbf{31.081\,lbf·ft}$$

3.2.2 THE ENGLISH GRAVITATIONAL SYSTEM

Different from the English Engineering System, in the English Gravitational System, the force unit lbf and the acceleration unit ft/s^2 are defined first; then, the mass unit is determined according to Equation (3.1),

$$m = \frac{F}{a}$$

In the English Gravitational System,

$$\text{Units of } m = \frac{\text{lbf}}{\dfrac{\text{ft}}{\text{s}^2}} = \frac{\text{lbf·s}^2}{\text{ft}}$$

the mass unit lbf·s^2/ft is conventionally defined as a slug:

$$1\ \text{slug} = 1\frac{\text{lbf·s}^2}{\text{ft}} \tag{3.7}$$

The slug unit is a unit combination of lbf and s^2/ft. In Equation (3.1), the units of the force in unit lbf, the mass in unit slug, and the acceleration in unit ft/s^2 are consistent in the English gravitational system, that is,

$$F\ \text{lbf} = (m\ \text{slug})\left(a\frac{\text{ft}}{\text{s}^2}\right) \tag{3.8}$$

Unit conversion is not necessary in Equation (3.8). Both the unit slug in the English Gravitational System and the unit lbm in the English Engineering System are mass units. However, they are not the same. The relationship between the slug and lbm is

$$m \text{ slug} = \frac{m\,\text{lbm}}{g_c \dfrac{\text{lbm·ft}}{\text{lbf·s}^2}} \tag{3.9}$$

Exercise 3.6

Derive the relationship of mass units in slug and lbm expressed in Equation (3.9).

SOLUTION

Using Equation (3.2),

$$F\,\text{lbf} = \frac{(m\,\text{lbm})\left(a\,\dfrac{\text{ft}}{\text{s}^2}\right)}{g_c\dfrac{\text{lbm·ft}}{\text{lbf·s}^2}}$$

and Equation (3.8),

$$F\,\text{lbf} = (m\,\text{slug})\left(a\,\dfrac{\text{ft}}{\text{s}^2}\right)$$

and combining the two equations, the relationship of mass units in the slug and lbm, therefore, is obtained:

$$m \text{ slug} = \frac{\dfrac{(m\,\text{lbm})\left(a\,\dfrac{\text{ft}}{\text{s}^2}\right)}{g_c\dfrac{\text{lbm·ft}}{\text{lbf·s}^2}}}{a\,\dfrac{\text{ft}}{\text{s}^2}} = \frac{m\,\text{lbm}}{g_c\dfrac{\text{lbm·ft}}{\text{lbf·s}^2}}$$

which is Equation (3.9).

Exercise 3.7

Estimate (a) the mass unit in slug for a 1 lbm object in a specified gravitational field of 27.5 ft/s² and (b) the force generated by the object in the field by using the mass unit in lbm and slug, respectively.

SOLUTION

(a) Using Equation (3.9), the mass in slug is determined to be

$$m = \frac{m\,\text{lbm}}{g_c\dfrac{\text{lbm·ft}}{\text{lbf·s}^2}} = \frac{1\,\text{lbm}}{32.174\dfrac{\text{lbm·ft}}{\text{lbf·s}^2}} = 31.081 \times 10^{-3}\,\text{slug}$$

(b) Using Equation (3.2), the force generated by the object in the field is found as

$$F = \frac{(m\,\text{lbm})\left(a\dfrac{\text{ft}}{\text{s}^2}\right)}{g_c\dfrac{\text{lbm·ft}}{\text{lbf·s}^2}} = \frac{(1\,\text{lbm})\left(27.5\dfrac{\text{ft}}{\text{s}^2}\right)}{32.174\dfrac{\text{lbm·ft}}{\text{lbf·s}^2}} = 0.855\ \textbf{lbf.}$$

Using Equation (3.8), the force generated by the object in the field is found,

$$F\ \text{lbf} = (m\,\text{slug})\left(a\frac{\text{ft}}{\text{s}^2}\right)$$

$$= \left(31.081\times10^{-3}\,\text{slug}\right)\left(27.5\frac{\text{ft}}{\text{s}^2}\right) = 0.855\ \textbf{lbf}$$

The results are identical.

Exercise 3.8 (FE style)

Density unit can be expressed as that shown below.

(A) $\dfrac{\text{lbf}}{\text{ft}^3}$

(B) $\dfrac{\text{slug}}{\text{ft}^3}$

(C) $\dfrac{\text{lbm}}{\text{ft}^3}$

(D) $\dfrac{\text{lbm}}{\text{lbf}}$

SOLUTION

The correct answers are **(B)** and **(C)**. Both slug and lbm are mass units. (A) is the unit of specific weight, and (D) has no physics meaning.

3.2.3 THE ABSOLUTE ENGLISH SYSTEM

In the Absolute English System, the mass unit lbm and the acceleration unit ft/s² are defined first. The force unit is then determined from Newton's second law. The name of the force unit defined in the Absolute English System is known as a poundal, denoted as pdl, that is,

$$F\ \text{pdl} = (m\,\text{lbm})\left(a\frac{\text{ft}}{\text{s}^2}\right) \tag{3.10}$$

Therefore,

$$1\ \text{pdl} = 1\frac{\text{lbm·ft}}{\text{s}^2} \tag{3.11}$$

In the Absolute English System, the units of the force pdl, the mass lbm, and the acceleration ft/s² in Equation (3.10) are consistent. Therefore, a unit conversion is not necessary. It has to be pointed out that the poundal (pdl) is not the same as the pound-force (lbf) or the pound-mass (lbm). The poundal is a force unit in Absolute English System. It is a unit combination of lbm and ft/s².

Exercise 3.9

Estimate the force in poundal to accelerate an object at 10 ft/s² with a mass of 120 lbm.

SOLUTION

Referring to Equation (3.10), the force in poundal in the absolute English system is determined to be

$$F = (m \, \text{lbm})\left(a\frac{\text{ft}}{\text{s}^2}\right)$$

$$= (120 \, \text{lbm})\left(10\frac{\text{ft}}{\text{s}^2}\right) = 12{,}000\frac{\text{lbm·ft}}{\text{s}^2} = \mathbf{1{,}200 \, pdl}$$

Exercise 3.10 (FE style)

1 poundal is equal to

(A) $1\dfrac{\text{slug·ft}}{\text{s}^2}$.

(B) $32.174\dfrac{\text{slug·ft}}{\text{s}^2}$.

(C) $1\dfrac{\text{lbm·ft}}{\text{s}^2}$.

(D) $32.174\dfrac{\text{lbm·ft}}{\text{s}^2}$.

SOLUTION

The correct answer is **(C)**. Referring to Equation (3.11), $1\,\text{pdl} = 1\dfrac{\text{lbm·ft}}{\text{s}^2}$. (A) is the force unit in the English Gravitational System, $1\,\text{lbf} = 1\dfrac{\text{slug·ft}}{\text{s}^2}$. (B) has no physics meaning. (D) is the force unit in the English Engineering System, $1\,\text{lbf} = 32.174\dfrac{\text{lbm·ft}}{\text{s}^2}$.

3.3 THE SI

The International System of Units (abbreviated SI from the French, Système International d'Unités) is a globally accepted international unit system. The SI was created in 1960 by the General Conference of Weights and Measures (abbreviated CGPM

from the French, Conférence Générale des Poids et Mesures). The system was regu-
lated and is continually developed by three international organizations coordinately
that were established in 1875 in terms of the meter convention. The three organizations are the CGPM, the International Committee of Weights and Measures (abbreviated CIPM from the French, Comité International des Poids et Mesures), and the
International Bureau of Weights and Measures (abbreviated BIPM from the French,
Bureau International des Poids et Mesures; see A.1 in Appendix). The SI is a simple
and logical system based on a decimal relationship among units. There are two subsystems in the SI: One is the CGS (centimeter–gram–second) system, and another
is the MKS (meter–kilogram–second) system. The current SI units in engineering
practice are developed from the MKS units. The convenience of using SI units lies
in their inherent unit consistency and systematic prefixes for modifying the unit
magnitudes.

3.3.1 THE CGS SYSTEM

The CGS system formalized the concept of a coherent system of units in 1874. The
system was named for its three fundamental units of length, mass, and time: centimeter, gram, and second. They are symbolized as cm, g, and s, respectively. The CGS
system is used widely by chemists, physicists, and theoretical researchers.

When Newton's second law, shown in Equation (3.1), is written in the units of the
CGS system, the unit of force can be obtained from

$$F = (m\,g)\left(a\,\frac{cm}{s^2}\right)$$

$$= ma\,\frac{g\cdot cm}{s^2} = ma\,dyn$$

The derived units in the CGS system can be composed of the three fundamentals
units. For example, the unit of force (g.cm/s²) is a combination of g, cm, and s²
denoted as a dyne by the symbol "dyn". The unit of energy (g.cm²/s²) is a combination
of g, cm², or dyne, and cm is denoted as an ergon in a symbol "erg". The unit of power
then is erg/s (ergon per second). The unit of volume is a liter (1,000 cubic centimeter)
in a symbol "L". Table 3.3 shows the fundamental units in the CGS system and some
commonly used derived units.

Exercise 3.11

The Newton's law of gravitation is expressed as

$$F = G\frac{m_1 m_2}{r^2}$$

where F is the mutual force of attraction between two particles, G is the universal
constant of gravitation, m_1 and m_2 are the masses of the two particles, and r is the
distance between the center of two particles.
Determine the unit of the universal constant of gravitation G in the CGS system.

TABLE 3.3

Fundamental Units and Some Commonly Derived Units in the CGS System

Quantity	Unit Name	Symbol
Fundamental Units		
Length *l*	centimeter	cm
Mass *m*	gram	g
Time *t*	second	s
Derived Units		
Volume *V*	liter	L
Acceleration *a*	galileo	Gal
Dynamic viscosity μ	poise	P
Electric current *I*	biot	Bi
Energy *E*	ergon	erg
Force *F*	dyne	dyn
Heat energy *Q*	calorie	cal
Illumination I_x	phot	ph
Kinematic viscosity v	stokes	St
Magnetomotive force F_m	gilbert	Gb
Power *P*	ergon per second	erg/s
Pressure *P*	barye	ba
Velocity *V*	centimeter per second	cm/s

SOLUTION

Knowing that the units of *F*, *m*, and *r* are dyn, g, and cm in the CGS system, respectively, the unit of the universal constant of gravitation *G*, therefore, is determined to be

$$G = \frac{Fr^2}{m_1 m_2} = \frac{(F \text{ dyn})(r\,\text{cm})^2}{(m_1 \text{g})(m_2 \text{g})} = \frac{Fr^2}{m_1 m_2} \frac{\mathbf{dyn \cdot cm^2}}{\mathbf{g^2}}$$

3.3.2 THE MKS SYSTEM

Adopted in 1889, the application of the MKS system of units succeeded the CGS system in commerce and engineering to avoid the shortage of relatively small unit of measurement in the CGS units. The MKS system is named for its three fundamental units of length, mass, and time: meter, kilogram, and second. They are symbolized as m, kg, and s. The MKS system was endorsed in 1962 by the International Standard Organization (ISO) and the International Electrochemical Commission (IEC). The MKS system comprises a coherent system of units of measurement starting with seven fundamental units. In October 1971, a complete metric system of units

TABLE 3.4
Fundamental Units and Commonly Used Derived Units in the MKS System

Quantity	Unit Name	Symbol
Fundamental Units		
Length l	meter	m
Mass m	kilogram	kg
Time m t	second	s
Temperature T	kelvin	K
Electric current I	ampere	A
Luminous intensity I_v	candela	cd
Amount of substance n	mole	mol
Derived Units		
Area A	square meter	m²
Acceleration a	–	a
Dynamic viscosity μ	pascal second	kg/m.s
Density ρ	kilogram cubic meter	kg/m³
Energy E	joule	J
Work W	joule	J
Heat Q	joule	J
Force F	newton	N
Illumination I_x	lux	lx
Kinematic viscosity v	meter squared per second	m/s²
Magnetomotive force F_m	amp-turn	At
Power P	watt	W
Pressure P	pascal	Pa
Velocity V	meter per second	m/s
Volume V	cubic meter	m³

replacing the MKS system was and named the SI (French: Système International d'Unités). The SI is a fully unit consistent system with a logical decimal structure among units. The SI, the metric system, and the MKS system are generally identical. Table 3.4 shows the fundamental units in the MKS system and some commonly used derived units.

Just as the other systems, the SI treats Newton's second law as a basic equation. The unit of force is derived from Equation (3.1). The force unit name is newton, symbolized as N. The unit N is a unit combination of the mass units in kg and the acceleration unit in m/s², that is,

$$F = (m\,\text{kg})\left(a\frac{\text{m}}{\text{s}^2}\right) = ma\frac{\text{kg·m}}{\text{s}^2} = ma\,\text{N}$$

Logically, therefore, the energy unit in the SI is a combination of the force unit and the length unit and is symbolized as N.m, or its equivalent kg·m²/s², named a joule, symbolized as J. The power unit is a joule per second, called a watt and symbolized as W. More derived units commonly used in engineering applications are shown in A.2 in the Appendix.

Exercise 3.12

An object of mass m = 20 kg is raised to a height h = 2.5 m vertically by a fork truck using a time period t = 10 s. Neglecting any frictions, determine the power P consumed by the fork truck doing the work in the unit W.

SOLUTION

The potential energy ΔEp. obtained for the object and the power consumed by the fork truck are determined by equations

$$\Delta E_p = mg\Delta h \qquad\qquad\text{(E3.12a)}$$

and

$$P = \frac{\Delta E_p}{t} \qquad\qquad\text{(E3.12b)}$$

Substituting the values of the mass and the height into Equation (E3.12a), the potential energy obtained for the object is

$$\Delta E_p = (20\,\text{kg})\left(9.81\frac{\text{m}}{\text{s}^2}\right)(2.5\,\text{m})$$
$$= 549.36\,\frac{\text{kg·m}^2}{\text{s}^2}$$
$$= 549.36\,\text{J}$$

and the power consumed by the fork truck by using Equation (E3.12b), therefore, is

$$P = \frac{549.36\,\text{J}}{10\,\text{s}} = 54.94\frac{\text{J}}{\text{s}} = \textbf{54.94\,W}$$

Exercise 3.13

Transmitting energy by using a rotating shaft in a mechanical system as shown in Figure *Exercise 3.13* is a common application in engineering practice. In the system, the torque T applied to the rotator typically is a constant,

$$T = Fr \qquad\qquad\text{(E3.13a)}$$

where r is the radius of the rotator. The shaft work is then determined to be

$$W_{sh} = Fs$$

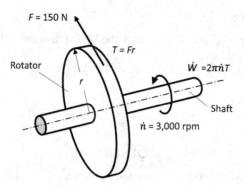

FIGURE *Exercise 3.13.*

where s is the rotation distance of the rotator. The expression s is as

$$s = (2\pi r)n$$

where n is the rotation revolutions of the rotator.

Then, the power transmitted through the shaft is decided by the shaft work done per unit time, which is expressed as

$$\dot{W}_{sh} = 2\pi \dot{n} T \qquad \text{(E3.13b)}$$

where $\dot{n}$ is the number of revolutions per unit time of the shaft.

If knowing that the radius of a rotator is 0.3 m, the force acting on the rotator is 150 N, and the shaft rotation rate is 3,000 revaluations per minute (rpm) as shown, determine the power transmitted through the shaft.

SOLUTION

Substituting the rotator radius $r = 0.3$ m and the acting force $F = 150$ N in Equation (E3.13a), the torque from the rotator is decided to be

$$T = Fr = (150\text{N})(0.3\text{m}) = 450 \text{ N·m}$$

Using the shaft rotation rate $\dot{n} = 3,000$ rpm, the shaft power is determined using Equation (E3.13b):

$$\dot{W}_{sh} = 2\pi \dot{n} T$$

$$= 2\pi\left(3,000\frac{1}{\text{min}}\right)\left(\frac{1\text{min}}{60\,\text{s}}\right)(450\,\text{N·m})$$

$$= 141.3\frac{\text{N·m}}{\text{s}} = 141.3\frac{\text{J}}{\text{s}}$$

$$= \textbf{141.3 W}$$

Exercise 3.14 (FE style)

2 meters, 4 kilograms, and 10 seconds in the CGS system are

(A) 2,000 mm, 4,000 g, 10 s.
(B) 2 m, 4,000 g, 10 s.
(C) 200 cm, 4 kg, 10 s.
(D) 200 cm, 4,000 g, 10 s.

SOLUTION

The correct answer is **(D)**. The CGS system is centimeter (cm)–gram (g)–second (s).

The units between the CGS system and MKS system can be converted. Table 3.5 shows some equivalent units between the systems. The format for converting the CGS unit to the MKS unit is to use the conversion factor shown in the table, that is,

$$\text{MKS unit} \times \text{Factor} = \text{CGS unit}$$

TABLE 3.5
Some Equivalent Units between the CGS System and the MKS System

Quantity		CGS		MKS		
Name	Symbol	Name	Unit	Name	Unit	Factor
Acceleration	a	Galileo	Gal	Meter per second squared	m.s^{-2}	0.01
dynamic viscosity	μ	Poise	P	Pascal second	Pa.s	0.1
Electric charge	Q	Franklin	Fr	Coulomb	C	3.34×10^{-10}
Electric current	I	Biot	Bi	Ampere	A	10
Electric dipole moment	μ	Debye	D	Coulomb meter	C.m	3.34×10^{-30}
Work	W	Erg	Erg	Joule	J	10^{-7}
Force	F	Dyne	Dyn	Newton	N	10^{-5}
Heat	Q	Calorie	Cal	Joule	J	4.187
Heat transmission	Q	Langley	Ly	Kilojoule per square meter	kJ·m^{-2}	41.84
Illumination	C	Phot	Ph	Lux	Lx	10^4
Kinematic viscosity	v	Stokes	St	Square meters per second	m^2·s^{-1}	10^{-4}
Magnetic field strength	H	Oersted	Oe	Ampere per meter	A·m^{-1}	79.577
magnetic flux	ϕ	Maxwell	Mx	Weber	Wb	10^{-8}
Magnetic flux density	B	Gauss	G	Tesla	T	10^{-4}
Magnetomotive force	F_m	Gilbert	Gi	Ampere	A	0.796
Pressure	P	Barye	Ba	Pascal	Pa	0.1
Wave number	v	Kayser	K	Per meter	m^{-1}	100

Exercise 3.15

A 20 kg slider is mounted on a horizontal rail. The slider moves along the rail by the action of a constant force $F = 180$ N from a cable as shown in Figure *Exercise 3.15*. The slider is released from the rest at the position A with an attached spring

having an extended initial length $x_1 = 0.15$ m. If neglecting all frictions and knowing the spring has a stiffness $k = 45$ N/m, determine the velocity V of the slider
using (a) the MKS system and (b) the CGS system as the slider reaches position B.

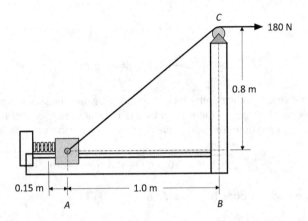

FIGURE *Exercise 3.15.*

SOLUTION

It can be seen that the stiffness of the spring is small enough to allow the spider
toward position B. A free-body diagram along x direction is drawn as shown.

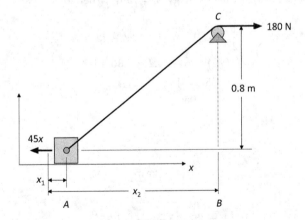

Knowing that the spring force 45x and the tension 180 N are the only forces exert-
ing on the system of the slider and the cable without considering any other frictions, so that

(a) Using the KGS system, as the slider moves from $x_1 = 0.15$ m to the position B,
the total moving distance of the slider is

$$0.15 \text{ m} = 1.0 \text{ m} = 1.15 \text{ m}$$

which is x_2. Therefore, the work W done by the spring force acting on the slider is found to be

$$
\begin{aligned}
W_{1-2} &= \frac{1}{2}k\left(x_1^2 - x_2^2\right) \\
&= \frac{1}{2}\left(45\frac{N}{m}\right)\left[(0.15\,m)^2 - (1.15\,m)^2\right] \\
&= \left(22.5\frac{N}{m}\right)(1.3\,m^2) = -29.25\,J
\end{aligned}
$$

The negative sign indicates the spring work acts against the slider moving direction. The work done on the system by the constant force of 180 N from the cable is the force times the horizontal displacement l of the cable over pulley C:

$$
l = \sqrt{(1.0\ m)^2 + (0.8\ m)^2} - 0.8\,m = 0.48\,m
$$

Thus, the work W_L done by the cable is

$$
W_L = (180\,N)(0.48\ m) = 86.4\ J
$$

By using the energy conservation principle, the work–energy equation to the system is

$$
T_2 - T_1 = W_{1-2} + W_L
$$

where T is the total kinetic energy, $T = \frac{1}{2}mv^2$. Therefore,

$$
\begin{aligned}
\frac{1}{2}(20kg)V^2 &= W_{1-2} + W_L + T_1 \\
(10kg)V^2 &= -29.25\,J + 86.4\ J + 0 \\
V^2 &= \frac{57.15\,J}{10\,kg} = 5.715\frac{J}{kg}
\end{aligned}
$$

Since

$$
1J = 1\frac{kg\cdot m^2}{s^2},
$$

$$
V = \sqrt{5.715\frac{kg\cdot m^2}{kg\cdot s^2}} = 2.39\frac{m}{s}
$$

(b) Using the CGS system, length is centimeter (cm), mass is gram (g), and time is second (s). All MKS units are converted to the CGS units. As the slider moves from $x_1 = 15$ cm to position B, the total moving distance of the slider is

$$
15\ cm + 100\ cm = 115\ cm
$$

which is x_2. The spring stiffness is $k = 0.45$ N/cm. Therefore, the work W done by the spring force acting on the slider is found to be

$$W_{1-2} = \frac{1}{2}k\left(x_1^2 - x_2^2\right)$$

$$= \frac{1}{2}\left(0.45\frac{N}{cm}\right)\left[\left(15\,cm\right)^2 - \left(115\,cm\right)^2\right]$$

$$= \left(0.225\frac{N}{cm}\right)\left(13,000\,cm^2\right) = -2,925\,N\cdot cm$$

The work done on the system by the constant force of 180 N from the cable is the force times the horizontal displacement l of the cable over the pulley C:

$$l = \sqrt{(100\ cm)^2 + (80\ cm)^2} - 80\ cm = 48\ cm$$

Thus, the work W_L done by the cable is

$$W_L = (180\,N)(48\,cm) = 8,640\,N\cdot cm$$

The work–energy equation to the system is

$$T_2 - T_1 = W_{1-2} + W_L$$

that is,

$$\frac{1}{2}(20,000\,g)V^2 = W_{1-2} + W_L + T_1$$

$$(1,000\,g)V^2 = -2,925\,N\cdot cm + 8,640\,N\cdot cm + 0$$

$$V^2 = \frac{5,715\,N\cdot cm}{10,000\,g} = 0.5715\frac{N\cdot cm}{g}$$

Since

$$1N = \frac{1,000\ g.100\ cm}{s^2}$$

$$V = \sqrt{57150\frac{g\cdot cm^2}{g\cdot s^2}} = 239\frac{cm}{s}$$

Alternatively, when (a) is completed, the velocity in the CGS system can be obtained directly by converting the result from (a):

$$V = 2.39\frac{m}{s} = 239\frac{cm}{s}$$

3.4   GUIDELINES FOR USING SI UNITS

The units in the SI are consistent. In addition to having the standardized fundamental units and derived units, there is a standardized syntax for expressing and writing SI units, symbols, and their formats. Some of the syntax might be known by common sense. However, some of the syntax may need to comply with conventional rules. As guidelines, commonly used rules for the SI units applied in engineering follow. These rules should be constantly consulted for correctly expressing and writing the units, symbols, and syntax. Most of the rules are applicable to the English system as well.

3.4.1   THE COMMON RULES

1. General

   - Only SI units and those recognized for use with the SI should be used. The equivalent units in the English system can be presented in parentheses after the SI units, for example,
     - The wall thickness of DN 50 (NPS 2) schedule 40 steel pipe is 3.81 mm (0.15 in).
   - To express units of measurement, the units of the quantities always should be assigned to numerical values, for example,

Quantity	Unit	Symbol
length	meter	m

     - The length of the road is about 3,000 m.
   - If a derived unit has a unique symbol, it is preferable to use the unique symbol instead of a combination of the fundamental units or others, for example,

     Improper:

     - The force acting on the table is 100 $kg.m/s^2$.
     - The pressure in the water is 1,200 N/m^2.
     - The energy rate consumed by the light bulb is about 30 kJ/s.

     Proper:

     - The force acting on the table is 100 N.
     - The pressure in water is 1,200 Pa or 1.2 kPa.
     - The energy rate consumed by the light bulb is about 30 W.

   - It is preferable to use standard units in calculation, such as length in m, mass in kg, force in N, and pressure in Pa (kPa or MPa is commonly used if Pa is too small). Nonstandard units are better to be converted to the standard units, for example,
     - Knowing the surface area of a plate is 400 mm^2, the mass of the plate is 5,000 g, and the acceleration of the plate is 200 cm/s^2, find the force and pressure acting on the plate.

Given $m = 5{,}000$ m, $g = 5$ kg, $a = 200$ cm/s² $= 2$ m/s², and $A = 400$ mm² $= 0.0004$ m²,

Then, $F = ma = (5$ kg$)(2$ m/s²$) = 10$ kg.m/s² $= 10$ N,

$P = F/A = 10$ N$/0.0004$ m² $= 25{,}000$ N/m² $= 25{,}000$ Pa $= 25$ kPa.

- The value of a quantity should not be used more than one unit, for example,

Improper:

- $l = 10$ m 23 cm 4 mm.

Proper:

- $l = 10.234$ m.

- Except for the fundamental unit of kilogram (kg), using a prefix in the denominator of composite units should be avoided, for example,

Improper:

- The force per unit length is measured as $F = 100$ N/mm.
- The specific volume of this rare gas is 3 cm³/mg.

Proper:

- The force per unit length is measured as $F = 100$ kN/m.
- The specific volume of this rare gas is 3 m³/kg.

2. Unit Names and Symbols

- Unit names and symbols should be in lowercase. However, unit symbols can be capitalized if the unit is named from a person's name, for example,
  - The absolute temperature symbol of the fundamental unit named after Lord Kelvin.
  - The force symbol of the derived unit named after Sir Isaac Newton.
  - The electrical inductance symbol of the derived unit named after Joseph Henry.
  - The power symbol of the derived unit named after James Watt.
  - The electric current symbol of the derived unit named after André-Marie Ampère.
  - The pressure symbol of the derived unit named after Blaise Pascal.

Quantity	Unit Name	Unit Symbol
Temperature	kelvin	K
Force	newton	N
Electrical inductance	henry	H
Power	watt	W

Quantity	Unit Name	Unit Symbol
Electric Current	ampere	A
Pressure	pascal	P
Volume	liter	L

Note: The degree symbol "°" is not applied for K. °K is improper. Similarly, °R is improper for expressing absolute temperature in the English system. It should be R.

- Unit names and symbols should not be mixed. Mathematical operations are not applied to unit names, for example,

  Improper:

  - $X = 10$ kilogram/m^3.
  - $Y = 10$ kg/cubic meter.
  - $Z = 10$ kilogram/cubic centimeter.
  - The density is 10 kg per m^3.

  Proper:

  - $X = 10$ kg/m^3.
  - $Y = 10$ kg.m^{-3}.
  - $Z = 10^7$ kg/m^3.
  - The density is 10 kg/m^3.
  - The density is 10 kilograms per cubic meter.

- No alterations nor changes to capitalization of unit symbols are allowed. For example, M does not equal m, and K does not equal k. There are no such units as Kg or MM.

  Improper:

  - $M = 6$ Kg.
  - The radius is 60 MM.

  Proper:

  - $m = 5$ kg.
  - The radius is 60 mm.

3. Numerals and Unit Symbols

- Numerical values and units of quantities should be expressed in acceptable format using Arabic numerals and symbols for units, for example,

  Improper:

  - $m =$ six kilograms.
  - $m =$ twenty kg.
  - $I = 18$ amperes.

Proper:

- $m = 6$ kg.
- $m = 20$ kg.
- $I = 18$ A.

4. Numbers and Numerical Values

- Numbers beginning a sentence must be spelled out. A hyphen should be used for 21 to 99 when using a number to start a sentence. The rule can be avoided by changing the order of words in a sentence so that the number is not in the initial position, for example,

Improper:

- 14 cups are on the table.
- 20 people are in the meeting.
- 31 people showed up at the conference.

Proper:

- Fourteen cups are on the table.
- There are 14 cups on the table.
- Twenty people are in the meeting.
- Thirty-one people showed up to the conference.

- The number 1 should be always spelled out in the sentence, for example,

Improper:

- The work lasted more than 1 hour.
- The operation lasted 1 day.
- 10 were successful among eleven tests, but the last 1 was not.

Proper:

- The work lasted more than one hour.
- The operation lasted one day.
- Ten were successful among 11 tests, but the last one was not.

- Whenever a numerical value is less than 1, a zero must always precede the decimal point. This avoids possible misinterpretation due to poor legibility of screens, fonts, prints, or photocopies, for example,

Improper:

- The disk diameter is .5 m.

Proper:

- The disk diameter is 0.5 m.

5. Abbreviations

- Only standard unit names, unit symbols, prefix symbols, and prefix names can be used. Improper use of abbreviations such as sec, cc, hrs, and mps should be avoided, for example,

Improper:

   − In experiments, they found the vessel velocity was 3 m/sec, the water density was 0.001 kg/cc, and the wind speed was 10 mps.
   − The steel example was tested at a temperature of 900 kelvins for 11 hrs.

Proper:

   − In experiments, they found the vessel velocity was 3 m/s, the water density was 1,000 kg/m^3, and the wind speed was 10 m/s.
   − The steel example was tested at a temperature of 900 K for 11 h.

- Unit symbols are never expressed without numerical values or in quantity symbols and not in abbreviations, for example,

Improper:

   − There are many m in a km.
   − This is sold by the m^3.
   − The rectangular yard is 10 *m* long and 6 *m* wide.

Proper:

   − There are 1,000 m in a km.
   − This is sold by the cubic meter.
   − The rectangular yard is 10 m long and 6 m wide.

6. Plurals and Periods

- Do not add an "s" after plural unit symbols. Unit symbols are unaltered in the plural, for example,

Improper:

   − 1 m = 100 cms.
   − 1 kg = 1,000 gs.

Proper:

   − 1 m = 100 cm.
   − 1 kg = 1,000 g.

- Unit names can be made plural when the numerical value that precedes them is more than one, for example,

Improper:

   − The volume of the smaller box is 0.25 liters.
   − It is about 250 milliliter.

Proper:

- The volume of the smaller box is 0.25 liter.
- It is about 250 milliliters.

• No periods need to be shown in unit symbols unless they appear at the end of a sentence, for example,

Improper:

- The plate area should be 2 m. long by 22 mm. wide.
- 2 km. long is enough for the wire connection.

Proper:

- The plate area should be 2 m long by 22 mm wide.
- Two km long is enough for the wire connection.

7. Multiplication and Division

• Two or more unit symbols should not be written together. A half-high dot is used to denote the multiplication of units. A slash or negative exponent is used to denote the division of units. No more than one slash can be used on the same unit line unless parentheses are used, for example,

Improper:

- The speed of sound is about 343 ms^{-1}.
- The specific volume of the liquid is 1.1348 m^3kg^{-1}.
- The heat transfer coefficient of the heat exchanger is up to 670 W/m^2K.
- The acceleration of the moving vehicle is over 15 m/s/s.
- The unit of this property is $N/s^2/m$.

Proper:

- The speed of sound is about 340 m/s or 340 $m.s^{-1}$.
- The specific volume of the liquid is 1.1348 m^3/kg or 1.1348 $m^3.kg^{-1}$.
- The heat transfer coefficient of the heat exchanger is up to 670 $W/m^2.K$.
- The acceleration of the moving vehicle is over 15 m/s^2 or 15 $m.s^{-2}$.
- The unit of this property is $N/(s^2/m)$.

8. Typeface

• Variables and quantity symbols should be shown in italic type. Numbers should be written in roman type and unit symbols are in roman type also, for example,

Improper:

- t = 7 s, where t is a variable symbol (time) and s is the unit symbol of time (second), respectively.
- T = 23 *K*, where T is a variable symbol (temperature) and *K* is the unit symbol of temperature (kelvin), respectively.

Proper:

- $t = 7$ s, where t is a variable symbol (time) and s is the unit symbol of time (second), respectively.
- $T = 23$ K, where T is a variable symbol (temperature) and K is the unit symbol of temperature (kelvin), respectively.

- Subscripts are in italic type if they represent variables, quantities, or running numbers. However, they should be in normal type if they are used for description, for example,

Type	Symbol	Unit Name
Variables	R_u	Universal gas constant
	c_p	Specific heat at constant pressure
	c_v	Specific heat at constant volume
	g_c	Gravitational constant
	v_g	Specific volume of gas
	u_f	Specific internal energy of fluid
Description	P_a	Absolute pressure
	P_{atm}	Atmospheric pressure
	V_{avg}	Average velocity
	T_{cr}	Critical temperature
	W_{spring}	Spring work
	E_{in}	Inlet energy
	V_{in}	Inlet velocity

9. Unit Modification and Percentages

- Unit symbols (or names) shouldn't be modified and mixed by addition of subscripts or other description, for example,

Improper:

- $V = 1{,}000\ V_{max}$.
- 10 % (by weight).
- The resistance is 100 Ω/square.

Proper:

- $V_{max} = 1{,}000 V$.
- A weight fraction of 10%.
- The resistance per square is 100 Ω.

- The symbol % is used simply to represent the number 0.01 and is always associated with a specific number, for example

Improper:

- The % of 0.003 is 0.3 %.

Proper:

- The percentage of 0.003 is 0.3%.

10. Information with Units

- Information should not be mixed with unit symbols or unit names, for example,

Improper:

- In the tank, the content is 350 mL H_2O/kg.
- In the tank, the content is 350 mL of water/kg.
- The length is 250 to 2,000 m.
- From 1 MHz–10 MHz or from 1 to 10 MHz.
- The frequency has a range of 8.2, 9.0, 9.5, 9.8, 10.0 GHz.
- From 20 °C–30 °C or from 20 to 30 °C.

Proper:

- In a tank, the water content is 350 mL/kg.
- The length is 250 m to 2,000 m.
- From 1 MHz to 10 MHz or (1 to 10) MHz.
- The frequency has a range of (8.2, 9.0, 9.5, 9.8, 10.0) GHz.
- From 20 °C to 30 °C or (20 to 30) °C.

11.   Math Notations

- It should be clear which unit symbol a numerical value belongs to and which mathematical operation applies to the value of a quantity, for example,

Improper:

- $Z = 35 \times 48$ cm.
- $G = 123 \pm 2$ g.
- $\Delta = 70 \pm 5\%$.

Proper:

- $Z = 35$ cm $\times 48$ cm.
- $G = 123$ g ± 2 g or (123 ± 2) g.
- $\Delta = 70\% \pm 5\%$ or $(70 \pm 5)\%$.

12. Time Units and Symbols

- The fundamental unit of time is second (s) in both the English system and the SI. It should be used in all technical calculations. But when time relates to calendar cycles, the time expressions of minute unit in min, hour unit in h, and day unit in d are acceptable (see A.5 in the Appendix), for example,
    - The arrival time must within 30 s.
    - One day has 24 hours, which means 1 d = 24 h.
    - The entire year has 365 days. Therefore, it is 1 y = 365 d.

13. Unit Spacing and Digit Spacing

- There should be a space between the numerical value and the unit symbol, for example,

    Improper:

    - In the room, there is a 25kg sphere.
    - The ladder is about 10-m high.
    - The wire is 8m long.

    Proper:

    - In the room, there is a 25 kg sphere.
    - The ladder is about 10 m high.
    - The wire is 8 m long.

- The temperature symbol °C, denoting degrees Celsius, should be preceded by a space between the temperature value and the symbol, which is the same for the absolute temperature K. But the symbol °C cannot be separated by a space, for example,

    Improper:

    - The room temperature is 22.4°C.
    - The room temperature is 38.2° C.
    - The fluid temperature is 300K.

    Proper:

    - The room temperature is 22.4 °C.
    - The room temperature is 38.2 °C.
    - The fluid temperature is 300 K.

- When the unit symbols for degree, minute, and second for plane angle, °, ′, and ″, respectively, it is not necessary to keep a space between the numerical value and the unit symbol, for example,

Improper:

- $\alpha = 30° 22' 8''$.
- $\alpha = 40° 12' 6''$.

Proper:

- $\alpha = 30°22'8''$.
- $\alpha = 40°12'6''$.

- When the value of a quantity is used as an adjective, a space should be left between the numerical value and the unit symbol. But if the spelled-out name of a unit is used, the normal rules of English can be applied, for example,

Improper:

- An 8-m height pole.
- A 11-kΩ resistor.
- A roll of 120 millimeter film.
- The thickness of insulation is 35-mm.

Proper:

- An 8 m height pole.
- An 11 kΩ resistor.
- A roll of 120-millimeter film.
- The thickness of insulation is 35 millimeters.

3.4.2 DECIMAL AND DIGIT GROUP SEPARATORS

Decimal Separators

A decimal separator also called a decimal sign, a decimal marker, or a decimal mark is a symbol used to separate the integer part from the fractional part of a number written in decimal form. Different countries designate the symbol type, either a period "." or a comma ",", to be used as the decimal separator. A period is commonly used in the United States and other English-speaking countries, and a comma is commonly used in continental Europe (see A.5 in the Appendix). For example, fifteen and one half can be expressed

using a comma: 15,5 mm (French).
using a period: 15.5 mm (the United States).

- The SI (ISO 80000–1:2022) stipulates that "The decimal sign is either a comma or a point on the line." The system does not stipulate a preference and observes that the type of the decimal separator depends on customary use in the language concerned but adds a note that per ISO/IEC directives, all ISO standards should use a comma as the decimal separator.

- The 22nd General Conference on Weights and Measures in 2003 declared that "the symbol for the decimal marker shall be either the period on the line or the comma on the line".

In other words, both the comma and the period are generally accepted as decimal separators for international use as shown in Figure 3.1. This book uses a period as the decimal separator except when there is a particular indication.

The choice of the symbol for the decimal separator also affects the choice of the symbol for the thousands separator used in digit grouping.

Digit Group Separators

When the digits of numerical values have more than four digits on the left side of the decimal separator the values are commonly separated into groups of three (thousands) by a symbol to facilitate reading. The symbol is called digit group separator also known as a digit group sign, digit group marker, or digit group mark. The convention for digit group separators varied from countries just as the decimal separators. The period "." is commonly used in the United States and other English-speaking countries. The comma "," is commonly used in continental Europe (see A.5 in the Appendix). For example, five thousand and five hundred is expressed as

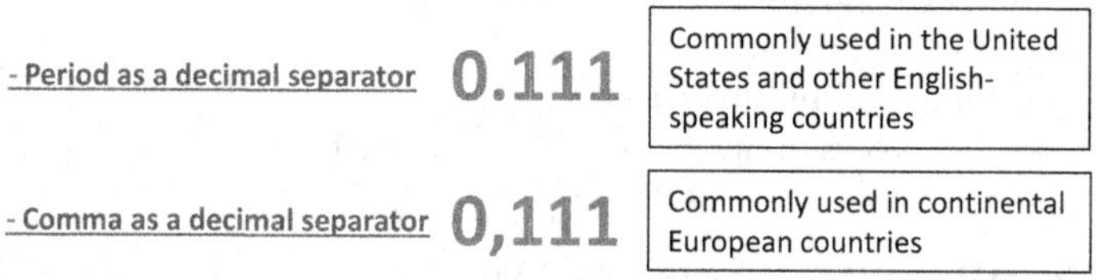

| - Period as a decimal separator | 0.111 | Commonly used in the United States and other English-speaking countries |
| - Comma as a decimal separator | 0,111 | Commonly used in continental European countries |

FIGURE 3.1   Usage of decimal separators.

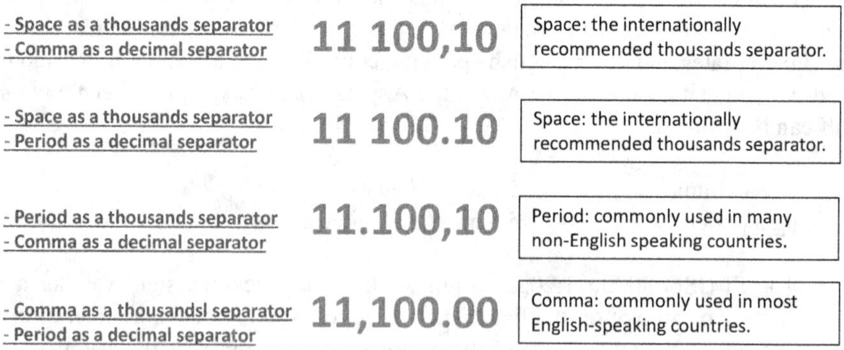

- Space as a thousands separator / - Comma as a decimal separator	11 100,10	Space: the internationally recommended thousands separator.
- Space as a thousands separator / - Period as a decimal separator	11 100.10	Space: the internationally recommended thousands separator.
- Period as a thousands separator / - Comma as a decimal separator	11.100,10	Period: commonly used in many non-English speaking countries.
- Comma as a thousandsl separator / - Period as a decimal separator	11,100.00	Comma: commonly used in most English-speaking countries.

FIGURE 3.2   Formats of digital group separators with decimal separators.

using a period: 5.500,00 m (French)
using a comma: 5,500.00 m (the United States).

Because of the confusion that could result in international documents, the SI unit system and ISO 80000–1:2022 advocated using a space as a digit group separator for a group of three. Therefore, there are three ways to group the number of more than ten thousand:

1. Space, the internationally recommended thousands separator
2. Period, the thousands separator used in many non-English speaking countries
3. Comma, the thousands separator used in most English-speaking countries

The mentioned digital group separators are used with the decimal separators to form four formats as shown in Figure 3.2.

As they can be seen, the period "." and the comma "," are the pair used for decimal separators and digital group separators, except for a space, which is used for the thousands separator recommended by the SI and ISO 80000–1:2022. Either one is with the other; that is, if "." is used as a decimal separator and another should be used as a digit group separator and vice versa, for example,

Improper:

– The value is 15559.012.
– The value is 15559,012.

Proper:

– The value is 15 559.012.
– The value is 15 559,012.
– The value is 15,559.012.
– The value is 15.559,012.

This book uses a comma "," as the digit group separator, such as "15,559.012" in the preceding example except when there is a particular indication.

3.4.3 UNIT PREFIXES

A unit prefix in the SI is a specifier or mnemonic prepended to units. The prefixes precede a basic unit of measure to indicate a decadic multiple and submultiple of a unit. Each prefix has a unique symbol that is prepended to the unit symbol. The unit prefixes are always considered to be part of the unit, for example, in exponentiation, so that 1 km^2 means one square kilometer, not 1,000 square meters, and 1 cm^3 means one cubic centimeter, not 100th of a cubic meter. Therefore, prefixes, added to a unit name, create the larger or smaller units as factors that are powers of 10. For example, add the prefix *kilo*, which means a thousand, to the unit gram to indicate 1,000 grams; thus, 1,000 grams become 1 kilogram. Table 3.6 shows the most commonly used unit prefixes in the SI.

TABLE 3.6
The Most Commonly Used Unit Prefixes in the SI

Numerical Value	Multiple	Thousands Base	Prefix	Symbol
1,000,000,000,000,000,000,000,000	10^{24}	$1,000^8$	yotta	Y
1,000,000,000,000,000,000,000	10^{21}	$1,000^7$	zetta	Z
1,000,000,000,000,000,000	10^{18}	$1,000^6$	exa	E
1,000,000,000,000,000	10^{15}	$1,000^5$	peta	P
1,000,000,000,000	10^{12}	$1,000^4$	tera	T
1,000,000,000	10^9	$1,000^3$	giga	G
1,000,000	10^6	$1,000^2$	mega	M
1,000	10^3	$1,000^1$	kilo	k
100	10^2		hecto	h
10	10^1		deca	da
1	10^0	$1,000^0$		
0.1	10^{-1}		deci	d
0.01	10^{-2}		centi	c
0.001	10^{-3}	$1,000^{-1}$	milli	m
0.000,001	10^{-6}	$1,000^{-2}$	micro	μ
0.000,000,001	10^{-9}	$1,000^{-3}$	nano	n
0.000,000,000,001	10^{-12}	$1,000^{-4}$	pico	p
0.000,000,000,000,001	10^{-15}	$1,000^{-5}$	femto	f
0.000,000,000,000,000,001	10^{-18}	$1,000^{-6}$	atto	a
0.000,000,000,000,000,000,001	10^{-21}	$1,000^{-7}$	zepto	z
0.000,000,000,000,000,000,000,001	10^{-24}	$1,000^{-8}$	yocto	y

Prefixes corresponding to powers of one thousand are usually preferred. However, units such as the hectopascal, hectare, decibel, centimeter, and centiliter are typically used too.

All the prefix symbols can also be formally made with upper- and lowercase Latin alphabet except for the symbol for micro, which is uniquely a Greek letter, "μ". The symbols for the prefixes as multiples are uppercase letters and the symbols for the prefixes as submultiples are lowercase letters with the exception of three: k (kilo), h (hecto) and da (deka). Table 3.7 presents the prefix application of power in watt.

The selection of the appropriate decimal multiple or submultiple of a unit for expressing the value of a quantity, that is, the choice of SI prefix, is governed by several factors. These include

- the need to indicate that digits of a numerical value are significant,
- the need to have numerical values that are easily understood, and
- the practice in a particular field of science or technology.

TABLE 3.7
SI Prefix Application of Power in Watt (W)

Submultiples			Multiples		
Value	Symbol	Name	Value	Symbol	Name
10^{-1} W	dW	deciwatt	10^{1} W	daW	decawatt
10^{-2} W	cW	centiwatt	10^{2} W	hW	hectowatt
10^{-3} W	**mW**	**milliwatt**	10^{3} W	**kW**	**kilowatt**
10^{-6} W	**μW**	**microwatt**	10^{6} W	**MW**	**megawatt**
10^{-9} W	**nW**	**nanowatt**	10^{9} W	**GW**	**gigawatt**
10^{-12} W	**pW**	**picowatt**	10^{12} W	**TW**	**terawatt**
10^{-15} W	**fW**	**femtowatt**	10^{15} W	**PW**	**petawatt**
10^{-18} W	aW	attowatt	10^{18} W	EW	exawatt
10^{-21} W	zW	zeptowatt	10^{21} W	ZW	zettawatt
10^{-24} W	yW	yoctowatt	10^{24} W	YW	yottawatt
10^{-27} W	rW	rontowatt	10^{27} W	RW	ronnawatt
10^{-30} W	qW	quectowatt	10^{30} W	QW	quettawatt

Note: Commonly used submultiples and multiples are in bold.

It is recommended for easily understanding large numerical values the prefix symbols should be chosen in such a way that numerical values are shown between 0.1 and 1,000 and the prefix symbols that represent the number 10 raised to a power. Conventionally, a multiple of 3 for the power is preferably used, for example,

- 4.3×10^{7} Hz may be written as 43×10^{6} Hz = 43 MHz
- 0.00732 g may be written as 7.32×10^{-3} g = 7.32 mg
- 5,801 W may be written as 5.801×10^{3} W = 5.801 kW
- 6.5×10^{-8} m may be written as 65×10^{-9} nm

If the digit of a numerical value of a quantity required to be expressed is significant, for example, the expression l = 1,200 m is not possible to tell whether the last two zeroes are significant. However, in the expression l = 1,200 km, which uses the SI prefix symbol of "k", the two zeroes are assumed to be significant because if they were not, the value of the quantity l would have been written l = 1.2 km (see Section 6.2 in Chapter 6).

There are some guidelines for common usage of the unit prefix in engineering applications:

- Compound prefix names or symbols are not permitted, for example:

Improper:

- The result is 100 mμm.
- The result is 50 Mkm.

Proper:

- The result is 100 nm.
- The result is 50 Gm.

- Prefixes are case-sensitive. Uppercase and lowercase letters have different meanings, for example,
  - mm is millimeter (one-thousandth of a meter), that is, 1 mm = 10^{-3} m = 0.001 m.
  - pm is picometer (one-trillionth of a meter), that is, 1 pm = 10−23 m = 0.000, 000,000,001 m.
  But,
  - Mm is megameter (one million meters), that is, 1 Mm = 10^6 m = 1,000,000 m.
  - Pm is petameter (one quadrillion meters), that is, 1 Pm = 10^{15} m = 1,000 ,000,000,000,000 m.

- The prefixes and their symbols are always prefixed to the unit symbols without intervening any space or period. This distinguishes a prefixed unit symbol from the product of unit symbols, for which a space or mid-height dot as a separator is required, for example,

Improper:

- The object speed is so fast. It only spends 100 m s to reach the destination.
- The object speed is so fast. It only spends 100 m·s. to reach the destination.

Proper:

- The object speed is so fast. It only spends 100 ms to reach the destination.

- Hyphens are not used between prefixes and unit names or between prefix symbols and unit symbols, for example,

Improper:

- 300 milli-gram, or 300 m-g.
- 245 kilo-meter, or 245 k-m.
- 1,000 tera-hertz, or 1,000 T-Hz.

Proper:

- 300 milligrams, or 300 mg.
- 245 kilometers, or 245 km.
- 1,000 terahertz, or 1,000 THz.

- Prefixes may not be used in combination with a single symbol. This includes the case of the base unit kilogram, which already contains a prefix, for example,

Improper:

– 1 microkilogram (µkg).

Proper:

– 1 milligram (mg).

• When a unit is raised to a power, a prefix symbol attached to the unit symbol is included in the power too, for example,

Improper:

– 1 km^2 = 10^3 mm^2.
– The area is 100 km^2 = 100 km·m.
– The volume of the tank is 1,000 mm^3 = 1,000 × 10^{-3} m^3.

Proper:

– 1 km^2 = 1 km × 1 km = 10^6 m^2.
– The area is 10 km^2 = 10(km)^2 = 10 (km × km).
– The volume of the tank is 1,000 mm^3 = 1,000 × 10^{-9} m^3 = 10^{-6} m^3.

4 Unit Conversions

Basically, the choice of a mass unit is the major factor in determining which unit system will be used in calculations. The standard units of mass are gram, pound-mass, kilogram, and slug involved in the English system and the SI as shown in Figure 4.1. They merely represent the different quantities of substance.

Based on Newton's second law,

$$F = ma \qquad (3.1)$$

the force unit is a derived unit and is able to be decided once the mass unit is selected. The commonly used units of force corresponding to the mass units shown in Figure 4.1 are the dyne, poundal, newton, and pound-force as shown in Figure 4.2.

Other units of the quantities can be derived from the selected units of mass and force. For example, if the units of mass and force are pound-mass (lbm) and pound-force (lbf), respectively, it can be known the unit system is the English Engineering System and the pressure, work, and power in the system are lbf/in^2, lbf.ft, and blf.ft/s as well. If the units of mass and force are gram (g) and dyne (dyn), it can be known the unit system is the CGS system in the SI, and the pressure, work, and power in the system are dyn/cm^2, dyn·cm, and dyn·cm/s as well.

Unit conversions are necessary in the system or between the systems in engineering practice. Unit conversion is a process for altering the numerical values and units of measured quantities from one to another without changing the measurement of the physically existing quantity. For example, the unit of mass may need to be converted from gram (g) to kilogram (kg) in the SI, or the unit of force may need to be converted from newton (N) in the SI to the pound-force in the English engineering system. The surface area of a table may be properly expressed in feet, but the floor area of a room may be properly expressed in meters. To compare the areas of the table and room, converting the areas in the same units is necessary; that is, either the surface area of the table is converted to meters or the floor area of the room is converted to feet. Keeping terms or quantities in the same units in calculations is called unit consistency (see Section 6.1 in Chapter 6).

4.1 UNIT CONVERSION FACTOR

Unit consistency requires that each quantity expressed in terms in an equation calculation has the same unit. The unit conversion factor (UCF) is a unit equivalent used to alter one unit to another based on the mathematical and physical relations between the units. When processing a unit conversion, an appropriate UCF needs to

DOI: 10.1201/9781003508977-4

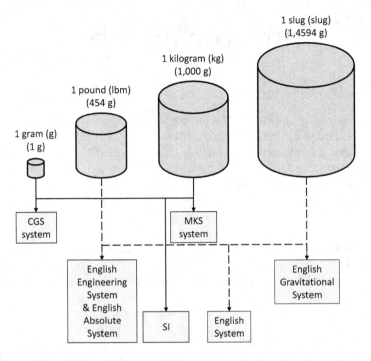

FIGURE 4.1   Units of mass.

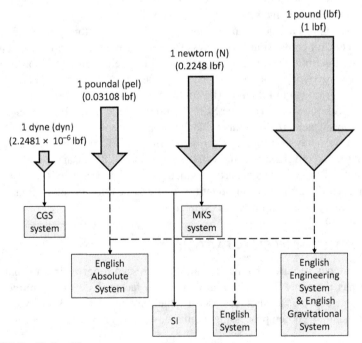

FIGURE 4.2   Units of force.

be selected. The process of unit conversion is by multiplication or division of the unit with the UCF:

$$\text{Original quantity } Q$$

$$= \text{Value of the original quantity N} \times \text{UCF} = \text{Designated Quantity } Q \quad (4.1a)$$

or

$$\text{Original quantity } Q$$

$$= \text{Value of the original quantity N/UCF} = \text{Designated Quantity } Q \quad (4.1b)$$

For example, Mr. John Smith's supervisor requires an urgent job to be completed in 3 hours 36 minutes. He lets John figure out the time in minutes to complete the job. John selects an appropriate unit conversion factor,

$$1\,h = 60\,min$$

Therefore,

$$3\,h = 3 \times 60\,min = 180\,min$$

As a result, the completion of the job by minutes is

$$3\,h\,36\,min = 180\,min + 36\,min = 216\,min$$

Minute is not a standard time unit in the SI. But it is acceptable to use with SI units (see A.5 in the Appendix).

Exercise 4.1

An isosceles triangle has a base length of 0.4 m and a side length of 500 mm as shown in Figure *Exercise 4.1*. Determine the area of the triangle in (a) square meters, (b) square millimeters, (c) square inches, and (d) square feet.

SOLUTION

Referring A.3b, the UCFs, the appropriate unit conversion factors of length are the following:

$$1\,mm = 0.001\,m$$
$$1\,mm = 0.0394\,in$$
$$1\,mm = 0.0033\,ft$$
$$1\,m = 1{,}000\,mm$$
$$1\,m = 39.370\,in$$
$$1\,m = 3.2808\,ft$$

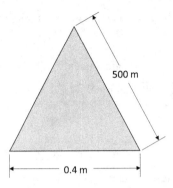

FIGURE *Exercise 4.1.*

(a) In meters

Side length: 500 mm = 500 × 0.001 m = 0.5 m
The height of the triangle, therefore, is

$$\text{Height} = (0.5\ m)\cos\left[\sin^{-1}\left(\frac{0.2m}{0.5m}\right)\right] = 0.4583\ m$$

The area of the triangle in square meters is determined to be

$$A = \frac{1}{2}(\text{Base length} \times \text{Height}) \quad = \frac{1}{2}(0.4\ m \times 0.4583\ m) = \mathbf{0.0917\ m^2}.$$

(b) In millimeters
Base length: 0.4 m = 0.4 × 1,000 mm = 400 mm
The height of the triangle, therefore, is

$$\text{Height} = (500\ mm)\cos\left[\sin^{-1}\left(\frac{200mm}{500mm}\right)\right] = 458.3\ mm$$

The area of the triangle in square millimeters is determined to be

$$A = \frac{1}{2}(\text{Base length} \times \text{Height})$$

$$= \frac{1}{2}(400\ m \times 458.3\ mm) = \mathbf{91{,}700\ mm^2}$$

(c) In inches
Base length: 0.4 × 39.370 in. = 15.75 in.
Side length: 500 mm = 500 × 0.0394 in. = 19.69 in.
The height of the triangle, therefore, is

$$\text{Height} = (19.69\ in)\cos\left[\sin^{-1}\left(\frac{7.87in}{19.69in}\right)\right] = 18.05\ in$$

The area of the triangle in square inches is determined to be

$$A = \frac{1}{2}(\text{Base length} \times \text{Height}) \quad = \frac{1}{2}(15.75\ in \times 18.05\ in) = \mathbf{142.14\ in^2}$$

(d) In feet
   Base length: 0.4 m = 0.4 × 3.2808 ft = 1.31 ft
   Side length: 500 mm = 500 × 0.00328 ft = 1.64 ft
   The height of the triangle, therefore, is

$$\text{Height} = \left(1.64\text{ ft}\right)\cos\left[\sin^{-1}\left(\frac{1.64\text{ft}}{1.64\text{ft}}\right)\right]=1.50\text{ ft}$$

The area of the triangle in square feet is determined to be

$$A = \frac{1}{2}(\text{Base length}\times\text{Height}) \quad =\frac{1}{2}(1.31\text{ ft}\times1.50\text{ ft})=\mathbf{0.98\text{ ft}^2}$$

Alternatively, the results of (b), (c), (d) can be obtained directly by using the result 0.0917 m² from (a) and the corresponding UCFs:

$$A = 0.0917\text{ m}^2 = 0.0917 \times 1{,}000^2\text{ mm}^2 = 91{,}700\text{ mm}^2$$

$$A = 0.0917\text{ m}^2 = 0.0917 \times 39.370^2\text{ in.}^2 = 142.14\text{ in.}^2$$

$$A = 0.0917\text{ m}^2 = 0.0917 \times 3.2808^2\text{ ft}^2 = 0.98\text{ ft}^2$$

4.2   UNITY CONVERSION RATIO

Similar to the UCF, the unity conversion ratio (UCR) is a unit equivalent used to convert one unit to another based on the mathematical and physical relations between the units. From the UCF, for example, the force units in the SI are

$$1\,\text{N}=1\,\text{kg}\frac{\text{m}}{\text{s}^2}$$

and in the English system

$$1\,\text{lbf}=32.174\,\text{lbm}\frac{\text{ft}}{\text{s}^2}$$

can be expressed as ratios, that is,

$$\frac{1\,\text{N}}{1\,\text{kg}\dfrac{\text{m}}{\text{s}^2}} = 1$$

and

$$\frac{1\,\text{lbf}}{32.174\,\text{lbm}\dfrac{\text{ft}}{\text{s}^2}} = 1$$

The ratio is denoted as the UCR. The UCR is unitless and numerically equal to 1. The UCR can be inserted conveniently into any equation calculations to properly convert units. Therefore, applying the UCF may be more convenient in equation calculation (see Section 6.1.2 in Chapter 6). The process of unit conversion is by multiplication or division of the unit with the UCR:

Original quantity Q

= Original quantity $Q \times$ UCR = Designated Quantity Q (4.2a)

or

Original quantity Q

= Original quantity Q/UCR = Designated Quantity Q.   (4.2b)

For example, the speedometer in a moving automobile on a highway indicates that the speed of the automobile is 100 km/h. The limited speed of the highway is 55 mil/h. Determine whether the automobile is speeding and what the speed of the motorcycle is in m/s.

Using the appropriate unity conversion ratio (see A.3c in the Appendix),

$$\frac{0.6214\,\text{mil}}{1\,\text{km}}$$

the automobile has the speed of

$$100\,\frac{\text{km}}{\text{h}} = 100\,\frac{\text{km}}{\text{h}} \times \frac{0.6214\,\text{mil}}{1\text{km}} = 62.14\,\frac{\text{mil}}{\text{h}}$$

Therefore, the automobile is speeding.

Using the appropriate unity conversion ratios,

$$\frac{1{,}000\text{m}}{1\text{km}}$$

$$\frac{3600\,\text{s}}{1\text{h}}$$

the automobile is at the speed

$$100\,\frac{\text{km}}{\text{h}} = 100\,\frac{\text{km}}{\text{h}} \times \frac{\dfrac{1{,}000\,\text{m}}{1\text{km}}}{\dfrac{3600\,\text{s}}{1\text{h}}} = 27.78\,\frac{\text{m}}{\text{s}}$$

Exercise 4.2

An airplane weighing 420,000 pounds is cruising at a speed of 245 knots at an altitude of 18,000 feet above the ocean. Determine the total energy of the airplane referring to the ocean in lbf.ft converted by using (a) the UCF and (b) the UCR.

SOLUTION

The units used in the calculation are in the English system. Knowing that a knot is the unit of nautical miles per hour and is equal to 1.69 feet per second, the UCF is

$$1 \text{ knot} = 1.69 \frac{\text{ft}}{\text{s}}$$

The speed of the airplane, therefore, is

$$245 \text{ knots} = 245 \times 1.69 \frac{\text{ft}}{\text{s}} = 414.05 \frac{\text{ft}}{\text{s}}$$

Using the equation of mass,

$$m = \frac{w \, g_c}{g}$$

the mass of the airplane is

$$m = \frac{(420,000 \text{ lbf}) \left(32.174 \frac{\text{lbmft}}{\text{lbfs}^2} \right)}{32.174 \frac{\text{ft}}{\text{s}^2}} = 420,000 \, \text{lbm}$$

The total energy of the airplane can be determined by

$$E = m \left(\frac{V^2}{2} + gh \right)$$

The total energy of the airplane is determined to be

$$E = m \left(\frac{V^2}{2} + gh \right)$$

$$= (420,000 \, \text{lbm}) \left[\frac{(414.05 \frac{\text{ft}}{\text{s}})^2}{2} + 32.174 \frac{\text{ft}}{\text{s}^2} \times 18,000 \text{ft} \right]$$

$$= (420,000 \, \text{lbm}) \left(664,850.7 \frac{\text{ft}^2}{\text{s}^2} \right)$$

$$= 2.7924 \times 10^{10} \frac{\text{lbm·ft}^2}{\text{s}^2}$$

Using the unit conversion factor,

$$1\frac{\text{lbm·ft}^2}{\text{s}^2} = 0.0311\,\text{lbf·ft}$$

$$2.7924\times10^{10}\,\frac{\text{lbm·ft}^2}{\text{s}^2} = 2.7924\times10^{10}\times0.0311\,\text{lbf·ft}$$

$$= 0.87\times10^9\,\text{lbf·ft}$$

The total energy of the airplane, therefore, is

$$E = \mathbf{0.87\times10^9\,lbf\text{·}ft}$$

The total energy of the airplane is

$$E = m(\frac{V^2}{2} + gh) = 2.7924\times10^{10}\,\frac{\text{lbm·ft}^2}{\text{s}^2}$$

Using the UCR,

$$\frac{32.174\text{lbm·}\dfrac{\text{ft}^2}{\text{s}^2}}{1\text{lbf·ft}},$$

the total energy of the airplane is determined to be

$$E = \frac{2.7924\times10^{10}\,\dfrac{\text{lbm·ft}^2}{\text{s}^2}}{\dfrac{32.175\text{lbm}\dfrac{\text{ft}^2}{\text{s}^2}}{1\text{lbf·ft}}} = 0.87\times10^9\;\textbf{lbf·ft}$$

4.3   CONVERSIONS IN THE ENGLISH SYSTEM

In the United States, the English system is widely used in engineering practice. It is known that the system is not consistent. Some commonly used UCF and UCR for unit conversions in the English system are shown in Table 4.1.

Exercise 4.3

The mass of total eggs in a bucket is 20 lbm, and the mass of the bucket itself is 5 lbm as shown in Figure *Exercise 4.3*. Determine the total weight in lbf of the eggs and the bucket.

TABLE 4.1
Some Commonly Used UCFs and UCRs in the English System

Dimension	Conversion	Unit Conversion Factor	Unity Conversion Ratio
Acceleration	ft/s^2 to in/s^2	$1\ ft/s^2 = 12\ in/s^2$	$\dfrac{12\ \frac{in}{s^2}}{1\ \frac{m}{s^2}}$
	ft/s^2 to mil/h^2	$1\ ft/s^2 = 0.6818\ mil/h^2$	$\dfrac{0.6818\ \frac{mil}{h^2}}{1\ \frac{ft}{s^2}}$
Area	square foot to square inches	$1\ ft^2 = 144\ in^2$	$\dfrac{144\ in^2}{1\ ft^2}$
	acre to square feet	$1\ acre = 43{,}560\ ft^2$	$\dfrac{43{,}560\ ft^2}{1\ acre}$
Force	lbf to $lbm \cdot ft/s^2$	$1\ lbf = 32.174\ lbm \cdot ft/s^2$	$\dfrac{32.174\ lbm \cdot \frac{ft}{s^2}}{1\ lbf}$
	kip to lbf	$1\ kip = 1{,}000\ lbf$	$\dfrac{1{,}000\ lbm}{1\ kip}$
Heat	Btu to $lbf \cdot ft$	$1\ Btu = 778.169\ lbf \cdot ft$	$\dfrac{778.169\ lbf \cdot ft}{1\ Btu}$
		$1\ therm = 100{,}000\ Btu$	$\dfrac{100{,}000\ Btu}{1\ therm}$
Length	foot to inches	$1\ ft = 12\ in$	$\dfrac{12\ in}{1\ ft}$
	mile to feet	$1\ mi = 5{,}280\ ft$	$\dfrac{5{,}280\ ft}{1\ mil}$
	foot to yards	$1\ ft = 0.3333\ yard$	$\dfrac{0.3333\ ft}{1\ ft}$
Mass	slug to pound-mass	$1\ slug = 32.174\ lbm$	$\dfrac{32.174\ lbm}{1\ slug}$
	pound-mass to ounces	$1\ lbm = 16\ oz$	$\dfrac{16\ oz}{1\ lbm}$
Power	hp to Btu/h	$1\ hp = 2{,}544.5\ Btu/h$	$\dfrac{2{,}544.5\ \frac{Btu}{h}}{1\ hp}$
	hp to $lbf \cdot ft/s$	$1\ hp = 550\ lbf \cdot ft/s$	$\dfrac{550\ \frac{lbf \cdot ft}{s}}{1\ hp}$
Pressure	atm to pound-force per square inches	$1\ atm = 14.696\ psi$	$\dfrac{14.696\ psi}{1\ atm}$
	psi to lbf/ft^2	$1\ psi = 144\ lbf/ft^2$	$\dfrac{14.696\ psi}{1\ atm}$

(Continued)

TABLE 4.1 (Continued)

Some Commonly Used UCFs and UCRs in the English System

Dimension	Conversion	Unit Conversion Factor	Unity Conversion Ratio
	psi to inch mercury	1 psi = 2.036 in Hg at 0 °C	$\dfrac{2.036 \text{ in Hg}}{1 \text{ psi}}$
	atm to inch mercury	1 atm = 29.92 in Hg at 0 °C	$\dfrac{29.92 \text{ in Hg}}{1 \text{ psi}}$
Velocity	mile per hour to feet per second	1 mil/h = 1.4667 ft/s	$\dfrac{1.4667 \frac{\text{ft}}{\text{s}}}{1 \frac{\text{mil}}{\text{h}}}$
	knot to feet per second	1 knot = 1.6878 ft/s	$\dfrac{1.6878 \frac{\text{ft}}{\text{s}}}{1 \text{ knot}}$
Volume	cubic foot to gallons	1 ft³ = 7.4805 gal	$\dfrac{7.4805 \text{ gal}}{1 \text{ ft}^3}$
	cubic foot to cubic inches	1 ft³ = 1,728 in³	$\dfrac{1,728 \text{ ft}^3}{1 \text{ ft}^3}$
Work	lbf·ft to lbm·ft²/s²	1 lbf·ft = 32.174 lbm·ft²/s²	$\dfrac{32.174\,\text{lbm}\cdot\frac{\text{ft}^2}{\text{s}^2}}{1\,\text{lbf}\cdot\text{ft}}$

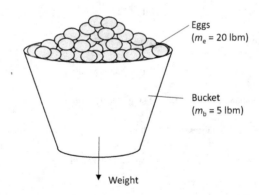

Eggs (m_e = 20 lbm)

Bucket (m_b = 5 lbm)

Weight

FIGURE *Exercise 4.3.*

SOLUTION

Referring to Table 4.1, the UCR is

$$\frac{1\,\text{lbf}}{32.174\,\text{lbm}\dfrac{\text{ft}}{\text{s}^2}}$$

Using the equation of weight

$$W = mg$$

the total weight of the eggs and the basket is determined to be

$$W = (20\text{lbm} + 5\text{lbm})\left(32.174\frac{\text{ft}}{\text{s}^2}\right)\left(\frac{1\text{lbf}}{32.174\text{lbm}\frac{\text{ft}}{\text{s}^2}}\right)$$

$$= \left(804.35\frac{\text{lbm·ft}}{\text{s}^2}\right)\left(\frac{1\text{lbf}}{32.174\text{lbm·}\frac{\text{ft}}{\text{s}^2}}\right) = 25\text{lbf}$$

Alternatively, using Equation (3.6a),

$$W = \frac{mg}{g_C}\text{lbf}$$

$$W = \frac{mg}{g_C} = \frac{(20\text{lbm} + 5\text{lbm})\left(32.174\frac{\text{ft}}{\text{s}^2}\right)}{32.174\frac{\text{lbm·ft}}{\text{lbf·s}^2}} = 25\text{lbf}$$

Exercise 4.4

A water hose with a nozzle is used to fill a 100 gal tank with water. The internal diameter of the hose is 3 in and is reduced to 2 in at the nozzle exit as shown in Figure *Exercise 4.4*. The tank is filled up by the water in 0.2 h. Knowing the water density is $\rho = 62.4$ lbm/ft^3 and neglecting all frictions, determine (a) the

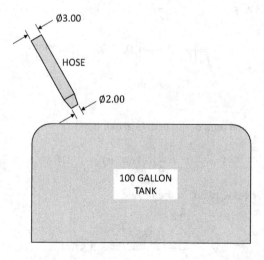

FIGURE *Exercise 4.4.*

volume mass flow in rate ft³/s and mass flow in rate lbm/s, (b) the water velocity in the unit of ft/s out of the nozzle, (c) the water velocity in the unit of ft/s in the hose.

SOLUTION

Referring to Table 4.1 and using the UCFs for all units in the fundamental units"

$$1\,\text{gal} = 0.1337\,\text{ft}^3$$
$$1\,\text{in} = 0.0833\,\text{ft}$$
$$1\,\text{h} = 3,600\,\text{s}$$

therefore, the tank volume is

$$100\,\text{gal} = 100 \times 0.1337\,\text{ft}^3 = 133.7\,\text{ft}^3$$

the nozzle diameter d at exist is

$$2\,\text{in.} = 2 \times 0.0833\,\text{ft} = 01.667\,\text{ft}$$

the hose internal diameter D is

$$3\,\text{in.} = 3 \times 0.0833\,\text{ft} = 0.2499\,\text{ft}$$

and the time period is

$$0.2\,\text{h} = 0.2 \times 3,600\,\text{s} = 720\text{s}$$

(a) The volume and mass flowrates of water are

$$\dot{V} = \frac{V}{\Delta t} = \frac{133.7\,\text{ft}^3}{720\text{s}} = \mathbf{0.1857}\,\frac{\mathbf{ft^3}}{\mathbf{s}}$$

$$\dot{m} = \rho\dot{V} = \left(62.4\,\frac{\text{lbm}}{\text{ft}^3}\right)\left(0.1857\,\frac{\text{ft}^3}{\text{s}}\right) = 11.59\,\frac{\mathbf{lbm}}{\mathbf{s}}$$

respectively.

(b) The flow sectional area of the nozzle at exit is

$$A_n = \frac{\pi}{4}d^2 = \frac{\pi}{4}(0.1667\,\text{ft})^2 = 0.218\,\text{ft}^2 ,$$

and the water velocity out of the nozzle is

$$V_n = \frac{\dot{V}}{A_n} = \frac{0.1857\,\dfrac{\text{ft}^3}{\text{s}}}{0.0218\,\text{ft}^2} = 8.5184\,\frac{\text{ft}}{\text{s}}$$

(c) The flow sectional area of the hose is

$$A_h = \frac{\pi}{4}D^2 = \frac{\pi}{4}(0.2499)^2 = 0.0491\,\text{ft}^2$$

and the water velocity out of the nozzle is

$$V_h = \frac{\dot{V}}{A_h} = \frac{0.1857\,\frac{\text{ft}^3}{\text{s}}}{0.0491\,\text{ft}^2} = 3.7821\frac{\text{ft}}{\text{s}}$$

Exercise 4.5 (FE style)

A kip means

(A) 1,000 psi.
(B) 1,000 lbf.
(C) 32,174 lbm·ft/s².
(D) 1,000 lbf·in.

SOLUTION

The correct answers are **(B)** and **(C)**. The word *kip* is a force unit that is an abbreviation of "kilopound-force", that is, 1,000 lbf. Also, referring to Table 4.1, 1 kip =1,000 lbf = 32,174 lbm·ft/s².

4.4 CONVERSIONS IN THE SI

The units in the SI, such as force N, mass kg, and length m, are consistent as per the Newton's second law. The system is based on a decimal relationship between the various units. The unit conversion is often easier within the SI due to the system's coherence and characteristics of metric prefixes that act as power-of-10 multipliers.

TABLE 4.2
Some Commonly Used UCF and UCR in the SI

Dimension	Conversion	Unit Conversion Factor	Unity Conversion Ratio
Acceleration	m/s² to cm/s²	1 m/s² = 100 cm/s²	$\dfrac{100\,\frac{\text{cm}}{\text{s}^2}}{1\,\frac{\text{m}}{\text{s}^2}}$
	km/s² to m/s²	1 km/s² = 1,000 m/s²	$\dfrac{1{,}000\,\frac{\text{m}}{\text{s}^2}}{1\,\frac{\text{km}}{\text{s}^2}}$

(Continued)

TABLE 4.2 (Continued)
Some Commonly Used UCF and UCR in the SI

Dimension	Conversion	Unit Conversion Factor	Unity Conversion Ratio
Area	square meter to square millimeters	$1 \text{ m}^2 = 1{,}000{,}000 \text{ mm}^2$	$\dfrac{1{,}000{,}000 \text{ mm}^2}{1 \text{ m}^2}$
	square kilometer to square meters	$1 \text{ km}^2 = 1{,}000{,}000 \text{ m}^2$	$\dfrac{1{,}000{,}000 \text{ m}^2}{1 \text{ km}^2}$
Force	N to kg.m/s²	$1 \text{ N} = 1 \text{ kg.m/s}^2$	$\dfrac{1 \text{ kg} \cdot \frac{m}{s^2}}{1 \text{ N}}$
	N to dyns	$1 \text{ N} = 100{,}000 \text{ dyn}$	$\dfrac{100{,}000 \text{ dyn}}{1 \text{ N}}$
Heat	kilojoule to joules	$1 \text{ kJ} = 1{,}000 \text{ J}$	$\dfrac{1{,}000 \text{ J}}{1 \text{ kJ}}$
	Joule to Newtown and meter	$1 \text{ J} = 1 \text{ N.m}$	$\dfrac{1 \text{ N·m}}{1 \text{ J}}$
Length	meter to millimeters	$1 \text{ m} = 1{,}000 \text{ mm}$	$\dfrac{1{,}000 \text{ mm}}{1 \text{ m}}$
	meter to centimeters	$1 \text{ m} = 100 \text{ cm}$	$\dfrac{100 \text{ cm}}{1 \text{ m}}$
	kilometer to meters	$1 \text{ km} = 1{,}000 \text{ m}$	$\dfrac{1{,}000 \text{ m}}{1 \text{ km}}$
Mass	kilogram to grams	$1 \text{ kg} = 1{,}000 \text{ g}$	$\dfrac{1{,}000 \text{ g}}{1 \text{ kg}}$
	metric tonnage to kilograms	$1 \text{ ton} = 1{,}000 \text{ kg}$	$\dfrac{1{,}000 \text{ kg}}{1 \text{ ton}}$
Power	watt to J/s	$1 \text{ watt} = 1 \text{ J/s}$	$\dfrac{1 \frac{J}{s}}{1 \text{ watt}}$
	kilowatt to watts	$1 \text{ kW} = 1{,}000 \text{ W}$	$\dfrac{1{,}000 \text{ W}}{1 \text{ kW}}$
Pressure	atm to kilopascals	$1 \text{ atm} = 101.325 \text{ kPa}$	$\dfrac{101.325 \text{ kPa}}{1 \text{ atm}}$
	kilopascal to pascals	$1 \text{ kPa} = 1{,}000 \text{ Pa}$	$\dfrac{1{,}000 \text{ Pa}}{1 \text{ kPa}}$
	pascal to N/m²	$1 \text{ Pa} = 1 \text{ N/m}^2$	$\dfrac{1 \text{ N}/\text{m}^2}{1 \text{ Pa}}$
	atm to millimeter mercury	$1 \text{ atm} = 760 \text{ mm Hg}$ @ 0 °C	$\dfrac{760 \text{ mm Hg}}{1 \text{ atm}}$

TABLE 4.2 (Continued)
Some Commonly Used UCF and UCR in the SI

Dimension	Conversion	Unit Conversion Factor	Unity Conversion Ratio
Velocity	meter per second to kilometers per hour	1 m/s = 3.60 km/h	$\dfrac{3.60\ \frac{km}{h}}{1\ \frac{m}{s}}$
Volume	cubic meter to liters	1 m³ = 1,000 L	$\dfrac{1,000\ L}{1\ m^3}$
	cubic meter to centimeters	1 m³ = 1,000,000 cm³	$\dfrac{1,000,000\ cm^3}{1\ m^3}$
Work	kilowatt and hour to kilojoules	1 kWh = 3,600 kJ	$\dfrac{3,600\ kJ}{1\ kWh}$

In the system, the unit conversions using the unit conversion factors and the unity conversion ratio are quite straightforward. Some commonly used unit conversion factors and unity conversion ratios for unit conversions in the SI are shown in Table 4.2.

Exercise 4.6

The gage pressure and temperature of air in a vehicle tire are measured P_1 = 250 kPa and T_1 = 22 °C before a trip and the gage pressure is P_2 = 260 kPa after the trip. Knowing the local atmospheric pressure is 1 atm and the tire volume is a constant, determine the air temperature T_2 in °C after the trip.

SOLUTION

Referring to Table 4.2 and using the unit conversion factor to convert atmospheric to the fundamental unit,

$$1 atm = 101.325 kPa$$

and

$$T(K) = T(^\circ C) + 273$$

then, the pressure P_1 and temperature T_1 in the tire before the trip is

$$P_1 = 250\ kPa + 101.325\ kPa = 351.325\ kPa$$
$$T_1 = 22\ ^\circ C + 273 = 295\ K$$

The pressure P_2 in the tire after the trip is

$$P_2 = 260\ \text{kPa} + 101.325\ \text{kPa} = 361.325\ \text{kPa}$$

Air is an ideal gas. Therefore, using the ideal-gas equation of state,

$$Pv = RT$$

the relationship of the properties before and after the trip is

$$\frac{P_1 V_1}{T_1} = \frac{P_2 V_2}{T_2}$$

The tire has a constant volume, $V_1 = V_2$. The temperature after the trip is determined to be

$$T_2 = \frac{P_2}{P_1} T_1 = \left(\frac{361.325\ \text{kPa}}{351.325\ \text{kPa}}\right)(295\text{K})$$
$$= 1.0285(295\text{K}) = 303.4\text{K} = \mathbf{30.4\ °C}$$

Exercise 4.7 (FE style)
The unit of 1,000 kg.m²/s² is equal to the following units.

(A) 1 kJ
(B) 1 kPa
(C) 1,000 N·m
(D) 1,000 N

SOLUTION

The correct answers are **(A)** and **(C)**. Using unit conversion factors,

$$1\text{kg.}\frac{m}{s^2} = 1\text{N}$$
$$1{,}000\ \text{J} = 1{,}000\ \text{N·m}$$
$$1{,}000\ \text{J} = 1\text{kJ}$$

therefore,

$$1{,}000\ \text{kg·m}^2/\text{s}^2 = 1{,}000\ \text{N·m} = 1{,}000\ \text{J} = 1\text{kJ}$$

4.5 CONVERSIONS BETWEEN THE SYSTEMS

The United States is a society that uses the dual-unit system. Some fundamental units and commonly used derived units in the SI and the English system are listed in Tables 4.3 and 4.4, respectively.

TABLE 4.3
Some Fundamental Units in the SI and the English System

| Quantity | SI Unit | | English Unit | |
	Name	Symbol	Name	Symbol
Mass m	kilogram	kg	pound	lbm
Length l	meter	m	foot	ft
Time t	second	s	second	s
Temperature T	kelvin	K	rankine	R

TABLE 4.4
Some Commonly Used Derived Units in the SI and the English System

| Quantity | SI Unit | | | English Unit | | |
	Name	Symbol	Unit	Name	Symbol	Unit
Force F	newton	N	$kg \cdot m/s^2$	pound-force	lbf	$32.174 \; lbm \cdot ft/s^2$
Pressure P	pascal	Pa	N/m^2	pound-force per square inch	psi	lb/in^2
Energy E	joule	J	$N \cdot m$	foot-pound	ft·lbf	$ft \cdot lbm2/s^2$
Heat Q	joule	J	$N \cdot m$	foot-pound	ft·lbf	$ft \cdot lbm2/s^2$
Work P	joule	J	$N \cdot m$	foot-pound	ft·lbf	$ft \cdot lbm2/s^2$
Power W	watt	W	J/s	horsepower	hp	$550 \; lbf \cdot ft/s$

Unit conversions between the English system and SI is common and frequently occur in engineering practice. Table 4.5 shows some commonly used unit conversions between the SI and the English system.

TABLE 4.5
Commonly Used Unit Conversions between the SI and the English System

Dimension	Unit Conversion	Unit Conversion Factor	Unity Conversion Ratio
Acceleration	m/s^2 to ft/s^2	1 m/s^2 to 3.2808 ft/s^2	$\dfrac{3.2808 \frac{ft}{s^2}}{1 \frac{m}{s^2}}$
Area	square meter to square feet	$1 \; m^2 = 10.764 \; ft^2$	$\dfrac{10.764 \; ft^2}{1 \; m^2}$
	acre to square meters	1 acre= 4,046 m^2	$\dfrac{4,046 \; m^2}{1 \; acre}$
Density	kg/m^3 to lbm/ft^3	$1 \; kg/m^3 = 0.062428 \; lbm/ft^3$	$\dfrac{0.062428 \frac{lbm}{ft^3}}{1 \frac{kg}{m^3}}$
Force	newton to pound-force	$1 \; N = 0.2248 \; lbf$	$\dfrac{0.2248 \, lbf}{1 \, N}$

(Continued)

TABLE 4.5   (Continued)
Commonly Used Unit Conversions between the SI and the English System

Dimension	Unit Conversion	Unit Conversion Factor	Unity Conversion Ratio
	kilogram-force to newton	1 kgf = 9.8067 N	$\dfrac{9.8067 \text{ N}}{1 \text{ kgf}}$
Heat	calorie to joules	1 cal = 4.184 J	$\dfrac{4.184 \text{ J}}{1 \text{ cal}}$
	kilojoule to British temperature unit	1 kJ = 0.9478 Btu	$\dfrac{0.9478 \text{ Btu}}{1 \text{ kJ}}$
	British temperature unit per pound to kilojoule per kilogram	1 Btu/lbm = 2.356 kJ/kg	$\dfrac{2.356 \frac{\text{kJ}}{\text{kg}}}{1 \frac{\text{Btu}}{\text{lbm}}}$
Length	meter to feet	1 m = 3.2808 ft	$\dfrac{3.2808 \text{ ft}}{1 \text{ m}}$
	centimeter to feet	1 cm = 0.032808 ft	$\dfrac{0.032808 \text{ ft}}{1 \text{ m}}$
	inch to millimeters	1 in = 25.4 mm	$\dfrac{25.4 \text{ mm}}{1 \text{ in}}$
Power	kilowatt to horsepower	1 kW = 1.341 hp	$\dfrac{1.341 \text{ hp}}{1 \text{ kW}}$
	kilowatt to Btu/h	1 kW = 3,412.14 Btu/h	$\dfrac{3{,}412.14 \frac{\text{Btu}}{\text{h}}}{1 \text{ kW}}$
	kilowatt to lbf.ft/s	1 kW = 737.56 lbf·ft/s	$\dfrac{737.56 \text{ lbf.} \frac{\text{ft}}{\text{s}}}{1 \text{ kW}}$
Pressure	atm to pressure per square inches	1 atm = 14.696 psi	$\dfrac{14.696 \text{ psi}}{1 \text{ atm}}$
	pressure per square inches to kilopascals	1 psi = 6.8948 kPa	$\dfrac{6.8948 \text{ kPa}}{1 \text{ psi}}$
	atm to inch mercury	1 atm = 29.92 in Hg @ 30 °C	$\dfrac{29.92 \text{ in Hg}}{1 \text{ atm}}$
	inch mercury to kilopascals	1 in Hg = 3.387 kPa	$\dfrac{3.387 \text{ kPa}}{1 \text{ in}}$
Velocity	meter per second to feet per second	1 m/s = 3.2808 ft/s	$\dfrac{3.2808 \text{ ft / s}}{1 \text{ m / s}}$
	mile per hour to kilometers per hour	1 mi/h = 1.6093 km/h	$\dfrac{1.6093 \frac{\text{km}}{\text{h}}}{1 \frac{\text{mi}}{\text{h}}}$
	knot to meters per second	1 knot = 0.5144 m/s	$\dfrac{0.5144 \frac{\text{ft}}{\text{s}}}{1 \text{ knot}}$
Volume	liter to gallons	1 L = 0.2642 gal	$\dfrac{0.2642 \text{ gal}}{1 \text{ L}}$

TABLE 4.5 (Continued)
Commonly Used Unit Conversions between the SI and the English System

Dimension	Unit Conversion	Unit Conversion Factor	Unity Conversion Ratio
	gallon to cubic meters	1 gal = 0.003485 m³	$\dfrac{0.003485\ m^3}{1\ gal}$
Work	Kilowatt and hour to British temperature unit	1 kWh = 3,412.14 Btu	$\dfrac{3,412.14\ Btu}{1\ kWh}$

Exercise 4.8

A U.S. manufacturing company is producing a piece of equipment for a customer in Europe. The piece of equipment weighs 1,220 lbf and is 9 feet long, 5.6 feet deep, and 6 feet high. In the information form, the weight W and volume V of the equipment need to be filled out in units of the SI. Please convert the units from the English system to the SI for (a) the weight and (b) the volume of the piece of equipment.

SOLUTION

In the SI, the standard units of weight and volume are N and m³. Referring to Table 4.5 and using the unity conversion ratios

$$\frac{4.4482\ N}{1\ lbf}$$

$$\frac{0.3048\ m}{1\ ft}$$

the units N and m³ of the equipment can be determined.

(a) The weight W of the equipment is

$$W = (1,220 \times lbf)\left(\frac{4.482\ N}{1\,lbf}\right) = 5,426.8\ N$$

(b) Since the length l, depth D, and height H are

$$l = 9\ ft\left(\frac{0.3048\ m}{1\,ft}\right) = 2.7432\ m$$

$$D = 5.6\ ft\left(\frac{0.3048\ m}{1\,ft}\right) = 1.7069\ m$$

$$H = 6\ ft\left(\frac{0.3048\ m}{1\,ft}\right) = 1.8288\ m$$

respectively, the volume V of the equipment is determined to be

$$V = (2.7432 \text{ m} \times 1.7069 \text{ m} \times 1.8288 \text{ m}) = \mathbf{8.5637 \text{ m}^3}$$

Exercise 4.9

A wind power plant with a rotor blade diameter of 10 m is installed at a location where the wind blows steadily at an average speed of 12 m/s and the air density is a constant $\rho = 1.19 \text{ kg/m}^3$ as shown in Figure *Exercise 4.9*. Neglecting any energy losses in the plant, determine the maximum power output that can be generated by the plant in (a) kW of the SI and (b) hp of the English system.

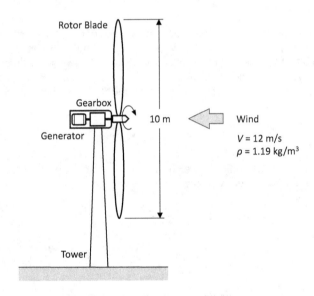

FIGURE *Exercise 4.9.*

SOLUTION

The maximum power output is obtained as the wind speed on the rotor blades is zero; that is, the surface of the rotor blade is a stagnation point. The kinetic energy per unit mass as the wind blows toward the rotor can be decided as

$$\text{ke} = \frac{V_1^2 - V_2^2}{2}$$

where V_1 and V_2 are the wind speeds toward and out from the rotor.
The maximum kinetic energy is

$$\text{ke}_{max} = \frac{V_1^2 - 0}{2} = \frac{V_1^2}{2}$$

Therefore, the maximum power output obtained from the blowing wind to the rotor is

$$P = \dot{m}(ke_{max}) = \dot{m}\frac{V_1^2}{2}$$

where $\dot{m}$ is the mass flow velocity of wind, $\dot{m} = \rho AV$ in which A is the surface area of the rotor $A = \pi D^2/4$.

(a) The maximum power output kW in the SI is determined to be

$$A = \frac{\pi D^2}{4} = \frac{3.14(10m)^2}{4} = 78.5\, m^2$$

$$\dot{m} = \rho AV = \left(1.19\frac{kg}{3^3}\right)\left(78.5\, m^2\right)\left(12\frac{m}{s}\right)$$

$$= 1,120.98\frac{kg}{s}$$

$$P = \dot{m}(ke_{max}) = \dot{m}\frac{V_1^2}{2}$$

$$= \frac{1}{2}\left(1,120.98\frac{kg}{s}\right)\left(12\frac{m}{s}\right)^2 = 80,710.56\frac{kg \cdot m^2}{s^3}$$

Referring to Table 4.3 and using the unity conversion ratios,

$$\frac{1,000\frac{m^2}{s^2}}{1\frac{kJ}{kg}}$$

$$\frac{1\, kW}{1\frac{kJ}{s}}$$

the maximum power is

$$P = \frac{80,710.56\frac{kg \cdot m^2}{s^3}}{1,000\frac{m^2}{s^2}} = 80.7\frac{kJ}{s}$$
$$\frac{}{1\frac{kJ}{kg}}$$

and becomes

$$P = 80.71\frac{kJ}{s} \times \frac{1\, kW}{1\frac{kJ}{s}} = \mathbf{80.7\ kW}$$

Converting all units of the SI to the units of the English system by using the unit conversion factors,

length: 1 m = 3.2808 ft

density: $1\dfrac{kg}{m^3} = 0.0624\dfrac{lbm}{ft^3}$

velocity: $1\dfrac{m}{s} = 3.2808\dfrac{ft}{s}$

then,

$$10\,m = 32.808\,ft$$

$$1.19\dfrac{kg}{m^3} = 0.0743\dfrac{lbm}{ft^3}\,;$$

$$12\dfrac{m}{s} = 39.37\dfrac{ft}{s}$$

the maximum power output can be determined to be

$$A = \frac{\pi D^2}{4} = \frac{3.14(32.808\,ft)^2}{4} = 844.9\,ft^4$$

$$\dot{m} = \rho AV = \left(0.0743\frac{lbm}{ft^3}\right)(844.9\,ft^2)\left(39.37\frac{ft}{s}\right)$$

$$= 2,471.5\frac{lbm}{s}$$

$$P = \dot{m}(ke_{max}) = \dot{m}\frac{V_1^2}{2g_c}$$

$$= \left(2,471.5\frac{lbm}{s}\right)\frac{\left(39.37\dfrac{ft}{s}\right)^2}{\left(2\left|32.174\dfrac{lbm\cdot ft}{lbf}\right|\dfrac{}{s^2}\right)}$$

$$= 59,532.8\frac{lbm\cdot ft^2}{s^3}$$

Using the UCR,

$$\frac{550\dfrac{lbm\cdot ft^2}{s^3}}{1hp}$$

the maximum power output is

$$P = \frac{59{,}532.8\dfrac{\text{lbm}\cdot\text{ft}^2}{s^3}}{550\dfrac{\dfrac{\text{lbm}\cdot\text{ft}^2}{s^3}}{1\text{hp}}} = 108.2 \text{ hp}$$

Alternatively, using the UCR,

$$\frac{1.341\,\text{hp}}{1\,\text{KW}}$$

to convert the result 80.7 kW from (a) directly, the identical maximum power output in hp can be obtained:

$$P = 80.7\,\text{kW} \times \frac{1.341\,\text{hp}}{1\,\text{KW}} = 108.2 \text{ hp}$$

Exercise 4.10

An ocean thermal energy conversion (OTEC) plant operating at a location where the ocean surface water temperature is 25 °C and the deep-water temperature under the surface is 5 °C as shown in Figure *Exercise 4.10*. Estimate the maximum thermal efficiency that the OTEC plant can obtain in the temperature difference ΔT calculated by using (a) the unit of the SI and (b) the unit of the English system.

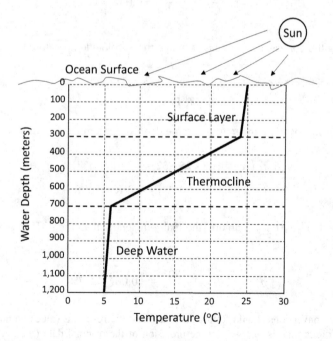

FIGURE *Exercise 4.10.*

SOLUTION

From thermodynamics, the maximum thermal efficiency that the OTEC plant can be obtained is determined by

$$\eta_{th} = 1 - \frac{T_L}{T_H}$$

where T_H and T_L are the absolute temperatures of the ocean surface and deep water, respectively.

(a) Using the conversion formula to obtain the absolute temperature K,

$$T(K) = T(°C) + 273,$$

the surface water temperature and the deep-water temperature in K are

$$T_H(K) = (25 + 273)K = 298K$$
$$T_H(K) = (5 + 273)K = 278K$$

respectively. The maximum thermal efficiency that the OTEC plant can be obtained is

$$\eta_{th} = 1 - \frac{T_H}{T_L} = 1 - \frac{278\ K}{298K} = 0.067 = \mathbf{6.7\%}$$

(b) Using the conversion formulas to obtain the absolute temperature R,

$$T(°F) = 1.8 \times T(°C) + 32$$
$$T(R) = T(°F) + 460$$

the surface water temperature and the deep-water temperature in R are

$$T_H(R) = [(1.8 \times 25 + 32) + 460]R = 537R$$
$$T_L(R) = [(1.8 \times 5 + 32) + 460]R = 501\ R$$

respectively. The maximum thermal efficiency that the OTEC plant can have is

$$\eta_{th} = 1 - \frac{T_H}{T_L} = 1 - \frac{501\ K}{537\ K} = 0.067 = \mathbf{6.7\%}$$

The unit conversions should follow the rules, and the resulting values do not use more significant digits than is justified by the precision of the original data (see Section 6.2 in Chapter 6), for example,

Improper:

- Converting a length of 34 inches to centimeters:

$$34 \text{ in} \times 2.54 \frac{\text{cm}}{\text{in}} = 86.36 \text{ cm}$$

- Converting a velocity of 30.2 meters per second to feet per second:

$$30.2 \frac{\text{m}}{\text{s}} \times \frac{3.2808 \frac{\text{ft}}{\text{s}}}{\frac{\text{m}}{\text{s}}} = 99.0802 \frac{\text{ft}}{\text{s}}$$

Proper:

- $$34 \text{ in} \times 2.54 \frac{\text{cm}}{\text{in}} = 86 \text{ cm}$$

$$30.2 \frac{\text{m}}{\text{s}} \times \frac{3.2808 \frac{\text{ft}}{\text{s}}}{\frac{\text{m}}{\text{s}}} = 99.1 \frac{\text{ft}}{\text{s}}$$

The resulting figure should be rounded to the minimum decimal place corresponding to those the involved terms possess in the unit conversion (see Section 6.3 in Chapter 6).

4.6 CONVERSIONS BY USING COMPUTER SOFTWARE

Software is a computer-based program. A software program for unit conversions is specially focused on converting units into the SI, the English system, and between the systems as well. A variety of unit conversion software programs are also called unit converters and are available on the internet. Many of them are free. Using unit conversion software is convenient and straightforward. Some of them can be downloaded, and some of them are only applicable online. Figures 4.3, 4.4, and 4.5 show the interfaces of some online unit converters.

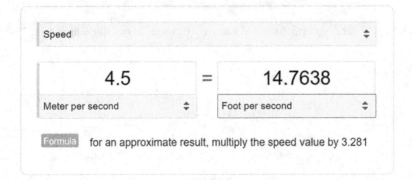

FIGURE 4.3 The converter from OnlineConversion. (Extracted from OnlineConversion. com. 2024.)

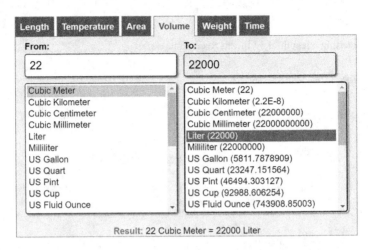

FIGURE 4.4   The converter from UnitConverters. (Extracted from UnitConverters.net. 2024.)

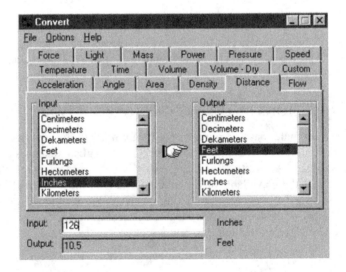

FIGURE 4.5   The converter from Joshmadison. (Extracted from Joshmadison.com. 2022.)

5 Dimensional Analysis

5.1 INTRODUCTION

Dimensional analysis is the study of relationships among different physical quantities. It is typically used for three prominent functions:

- To check for dimension homogeneity to keep unit consistency (see Section 2.3 in Chapter 2)
- To convert units correctly from one to another in the system or between the systems (see Chapter 4).
- To derive dimensionless members from physical phenomena and find the relations among quantities to guide design and R&D.

Checking for dimensional homogeneity to keep unit consistency and converting units are basic work in engineering practice. Dimensional analysis also serves as a powerful tool to derive dimensionless members also called dimensionless groups from physical phenomena and find the relations among quantities to guide design and R&D. A dimensionless number is a product or ratio of quantities without dimensions. The dimensionless numbers enable to

1. deepen the understanding of complex engineering phenomena.
2. reduce the amount of time needed to solve an equation numerically.

There are two methods commonly used to identify dimensionless numbers in the dimensional analysis. One is the method of the nondimensionalized equation, and the other method is the Pi theorem.

5.2 METHOD OF NONDIMENSIONALIZED EQUATION

The method of nondimensionalized equation is the nondimensionalization to remove the physical dimensions from equations involving physical quantities by a suitable procedure of variable substitution. For nondimensionalize equations, the following steps are commonly involved:

1. Identify all the independent and dependent variables.
2. Replace each of them with a quantity scaled relative to a characteristic unit of measure to be determined.
3. Divide through by the coefficient of the highest order polynomial or derivative term.
4. Choose judiciously the definition of the characteristic unit for each variable so that the coefficients of as many terms as possible become one.
5. Rewrite the equations in terms of their new dimensionless numbers.

DOI: 10.1201/9781003508977-5   **89**

Four typical equations in engineering,

- the continuity equation,
- the momentum equation,
- the energy equation, and
- the Navier–Stocks equation,

are used as examples to illustrate the method of nondimensionalized equations.

5.2.1 CONTINUITY EQUATION

The continuity equation of a steady one-dimensional flow is expressed as

$$\dot{m} = \rho V A = \text{constant} \tag{5.1}$$

Differentiating Equation (5.1), the equation is in a dimensionless form:

$$\frac{d\rho}{\rho} + \frac{dV}{V} + \frac{dA}{A} = 0 \tag{5.2}$$

Neglecting the potential energy, the conservation of energy for an isentropic flow without work interaction in differential form is

$$\frac{dP}{\rho} + V dV = 0 \tag{5.3}$$

Substituting Equation (5.3) into Equation (5.2) and rearranging, the equation becomes

$$\frac{dA}{A} = \frac{dP}{\rho} + \left(\frac{1}{V^2} - \frac{d\rho}{dP} \right) \tag{5.4}$$

Since

$$dP = a^2 d\rho \tag{5.5}$$

where a is the speed of sound. Combining Equations (5.4) and (5.5), the expression is

$$\frac{dA}{A} = \frac{dP}{\rho V^2} \left(1 - \text{Ma}^2 \right) \tag{5.6}$$

In Equation (5.6), Ma is defined as the Mach number,

$$\text{Ma} = \frac{V}{a} \tag{5.7}$$

Ma is a dimensionless number. It is a ratio of local velocity and the speed of sound. The Mach number Ma is an important parameter in the analysis of compressible fluid flow. When a fluid flows at the speed of Ma = 1, the flow is called a sonic flow.

Accordingly, when a fluid flows at the speed of Ma < 1 or Ma > 1, the flow is called a subsonic flow or supersonic flow, respectively. Figure 5.1 presents the flow types to the corresponding Ma. Table 5.1 shows the flow regimes to the corresponding Mach number.

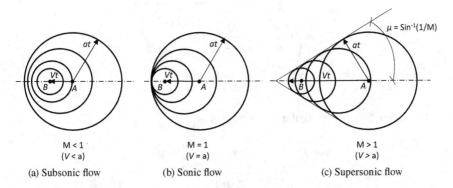

M < 1	M = 1	M > 1
(V < a)	(V = a)	(V > a)
(a) Subsonic flow	(b) Sonic flow	(c) Supersonic flow

FIGURE 5.1 Subsonic flow types to the corresponding Ma.

TABLE 5.1
Flow Regimes to the Corresponding Mach Number

Ma	< 0.8	0.8–1.0	1.0	1.0–1.3	1.3–5.0	5.0–10.0	>10.0
Regime	Subsonic	Transonic	Sonic	Transonic	Supersonic	Hypersonic	Hypervelocity

Exercise 5.1

Air entering a diffuser has a velocity of 220 m/s. Knowing the air temperature entering the diffuser is 32 °C as shown in Figure *Exercise 5.1*, determine (a) the Mach number of the air at the diffuser inlet and (b) the airflow region.

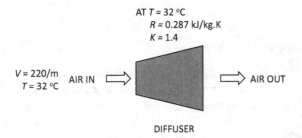

AT $T = 32\ °C$
$R = 0.287\ kJ/kg.K$
$K = 1.4$

$V = 220/m$
$T = 32\ °C$ AIR IN   ⟹   ⟹ AIR OUT

DIFFUSER

FIGURE *Exercise 5.1.*

SOLUTION

At an air temperature of 32 °C, the air constant R is 0.287 kJ/kg.K, and its specific heat ratio is 1.4. Then the speed of sound is determined to be

$$c = \sqrt{kRT}$$

$$= \sqrt{1.4\left(0.287\,\frac{kJ}{kg.K}\right)(305K)\left(\frac{1,000\,\frac{m^2}{s^2}}{1\,\frac{kJ}{kg}}\right)}$$

$$= 350.1\frac{m}{s}$$

(a) Using Equation (5.7), the Mach number is

$$Ma = \frac{V}{a} = \frac{220\,\frac{m}{s}}{350.1\,\frac{m}{s}} = 0.628 \cdot$$

(b) Referring to Table 5.1,

$$Ma < 0.8;$$

therefore, the airflow region is **subsonic**.

Exercise 5.2 (FE style)

When a flow is in the region of supersonic flow, the Ma of the flow is in the range of

(A) Ma = 1.0–1.3.
(B) Ma = 1.3–5.0.
(C) Ma = 5.0–10.0.
(D) Ma > 10.0.

SOLUTION

The correct answer is **(B)**. Referring to Table (5.1), a supersonic flow has Ma = 1.3–5.0.

5.2.2 MOMENTUM EQUATION

Figure 5.2 shows a hydraulic jump in an open-channel flow. If neglecting the surface tension and viscosity, the momentum equation describing the hydraulic jump with per unit width is expressed as

$$\frac{\gamma y_1^2}{2} - \frac{\gamma y_2^2}{2} = \frac{\gamma y_1 V_1}{g}(V_2 - V_1) \tag{5.8}$$

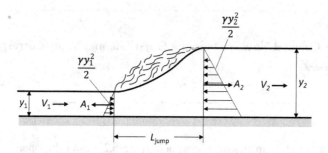

FIGURE 5.2 Hydraulic jump in an open-channel flow.

where y_1 and y_2 are the flow depths of upstream and downstream, respectively, $\{y\} = L$;

V_1 and V_2 are the flow velocities of upstream and downstream, respectively, $\{V\} = Lt^{-1}$;

γ is the specific weight of the fluid, $\{\gamma\} = ML^{-2}t^{-2}$; and

g is the gravitational acceleration, $\{g\} = Lt^{-2}$.

Rearranging Equation (5.8), then

$$\frac{\gamma y_1^2}{2}[1-(\frac{y_2}{y_1})^2] = V_1^2 \frac{\gamma}{g} y_1(1-\frac{y_2}{y_1})\frac{y_1}{y_2}. \tag{5.9}$$

Taking division by the term of $1 - y_2/y_1$ in Equation (5.9) and rearranging, the equation becomes a dimensionless equation:

$$\frac{V_1^2}{g y_1} = \frac{1}{2}\frac{y_2}{y_1}(1+\frac{y_2}{y_1}) \tag{5.10}$$

in which the left-hand side is a ratio of the inertia force to the gravitational force. The ratio is a dimensionless number defined as the Froude number Fr,

$$\text{Fr} = \frac{V}{\sqrt{gy}} \tag{5.11}$$

The equation to describe the hydraulic jump in the form involving the Fr becomes

$$\text{Fr}^2 = \frac{1}{2}\frac{y_2}{y_1}(1+\frac{y_2}{y_1})$$

In Equation (5.11), V is the average velocity of the flow, and y is the flow depth in the open channel. The open-channel flow can be classified as the subcritical, critical, and supercritical flow. Fr is an important dimensionless number that governs the

---

TABLE 5.2

The Open-Channel Flow Types in the Open Channel to the Corresponding Fr

Froude Number Fr	Fr < 1	Fr = 1	Fr > 1
Flow type	Subcritical flow	Critical flow	Supercritical flow

---

character of the flow in the open channel. Table 5.2 shows the open-channel flow types in the open channel to the corresponding Fr.

Considering the flow of a fluid in a rectangular open channel, the cross-sectional flow area is A_c. When the flow is critical, that is, Fr = 1, from Equation (5.11), the flow depth y is denoted as y_c. y_c is called the critical depth corresponding to Fr = 1 and determined as

$$y_c = \frac{\dot{V}^2}{gA_c^2}$$

For a rectangular open channel with a width b, the cross-sectional flow area is $A_c = by$, the critical depth is determined by

$$y_c = \left(\frac{\dot{V}^2}{gb^2} \right)^{\frac{1}{3}} \tag{5.12}$$

The flow types of an open-channel flow with the critical depth y_c are presented in Figure 5.3.

  (a) Definition of flows
  (b) Flow through a sluice gate

Exercise 5.3

Water in a rectangular open channel with a width $b = 0.5$ m flows steadily at a flow rate $\dot{V} = 300$ L³/s as shown in Figure *Exercise 5.3*. Knowing the flow depth $y = 22$ cm, determine (a) the flow velocity, (b) the Fr, (c) the critical depth y_c, and (d) the flow type.

SOLUTION

Referring to Table 4.2 and using the unit conversion factors to convert units to the fundamental units,

$$1 \frac{L^3}{s} = 0.001 \frac{m^3}{s}$$

$$1 \text{ cm} = 0.01 \text{ m}$$

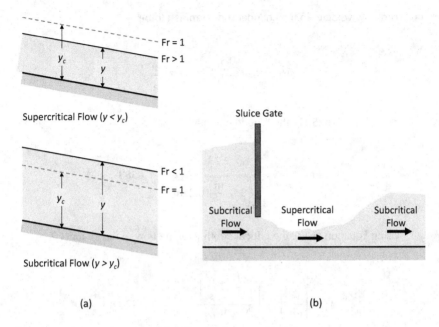

FIGURE 5.3 Critical flow, subcritical flow, and supercritical flow.

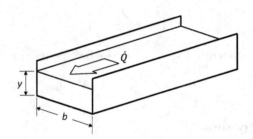

FIGURE *Exercise 5.3.*

the flow rate and depth in the channel are

$$\dot{V} = 300\frac{L^3}{s} = \left(300 \times 0.001\frac{m^3}{s}\right) = 0.3\frac{m^3}{s}$$

$$y = 22\ cm = 22 \times 0.01\ m = 0.22\ m$$

and the flow sectional area A_c is

$$A_c = yb = (0.22\ m)(0.5\ m) = 0.11\ m^2$$

(a) The flow velocity in the channel is determined to be

$$V = \frac{\dot{V}}{A_c} = \frac{0.3\frac{m^3}{s}}{0.11 m^2} = 2.73\frac{m}{s}$$

(b) Using Equation (5.11), the Froude number is

$$Fr = \frac{V}{\sqrt{gy}} = \frac{2.73\frac{m}{s}}{\sqrt{\left(9.81\frac{m}{s^2}\right)(0.22m)}} = 1.8583$$

(c) Using Equation (5.12), the critical depth y_c of the flow in the channel is

$$y_c = \left(\frac{\dot{V}^2}{gb^2}\right)^{\frac{1}{3}} = \left[\frac{\left(0.3\frac{m^3}{s}\right)^2}{(9.81\frac{m}{s^2})(0.5m)^2}\right]^{\frac{1}{3}} = 0.3323 \text{ m}$$

(d) Since the flow is

$$Fr = 1.8583 > 1$$

the flow is **supercritical**. It can be verified by using the critical depth y_c:

$$y = 0.22 \text{ m} < y_c = 0.3323 \text{ m}$$

5.2.3 ENERGY EQUATION

The Bernoulli equation is an energy-conservative equation. The equation describes a flow moving along its streamlines from one location to another location, the energy of the flow should be conservative. Figure 5.4 shows a flow along its streamlines in a pipe.

The energy of the flow from location 1 to location 2 is conservative and described by

$$\left(\frac{P}{\rho} + \frac{V^2}{2} + gz\right)_1 = \left(\frac{P}{\rho} + \frac{V^2}{2} + gz\right)_2 \tag{5.13}$$

or at any point, the energy of the flow shall be

$$\frac{P}{\rho} + \frac{V^2}{2} + gz = C \tag{5.14}$$

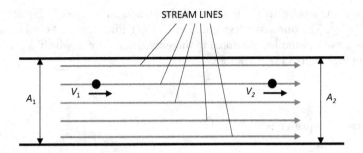

FIGURE 5.4 Flow streamlines in a pipe.

where P is the pressure, $\{P\} = ML^{-1}t^{-2}$;

V is the flow velocity, $\{V\} = Lt^{-1}$;
ρ is the fluid density, $\{\rho\} = ML^{-3}$;
z is the flow elevation, $\{z\} = L$; and
C is a constant.

Using scaling parameters, length L, velocity U_0, and density ρ_0 to nondimensionalize all variables in Equation (5.13),

$$P^* = \frac{P}{\rho_0 U_0^2}$$

$$\rho^* = \frac{\rho}{\rho_0}$$

$$V^* = \frac{V}{U_0}$$

$$g^* = \frac{gL}{U_0^2}$$

$$\rho^* = \frac{z}{L}$$

and substituting them in Equation (5.14), the equation becomes,

$$\rho_0 U_0^2 P^* + \frac{1}{2}\rho_0 \rho^* (U_0^2 V^{*^2}) + \rho_0 \rho^* g^* U_0^2 z^* = C \qquad (5.15)$$

Dividing by $\rho_0 U_0^2$ and assuming the flow is incompressible $\rho^* = 1$ in Equation (5.15), then

$$P^* + \frac{1}{2}V^{*^2} + g^* z^* = C^*$$

The nondimensionalization of the Bernoulli equation is obtained. In Equation (5.15), $C^* = C/\rho_0 U_0^2$. The dimensionless group gL/U_0^2 is defined as the Froud number Fr, which is a ratio of the inertial force to the gravitational force which is the same as shown in Equation (5.11). Therefore,

$$g^* = 1/Fr^2 \qquad (5.16a)$$

The Bernoulli equation with the Fr is

$$P^* + \frac{1}{2}V^{*2} + \frac{1}{Fr^2}z^* = C^* \qquad (5.16b)$$

Exercise 5.4 (FE style)

The Froude Number Fr is a dimensionless number measuring the ratio of

(A) Inertial force/Gravitational force.
(B) Gravitational force/Viscous force.
(C) Lift force/Dynamic force.
(D) Inertial force/Surface tension force.

SOLUTION

The correct answer is **(A)** described by Equation (5.11) and (5.16a).

5.2.4 THE NAVIER-STOKES EQUATIONS

The Navier–Stokes (N-S) equations are very useful because the equations help describe the mechanics of various engineering and scientific phenomena. They could be applied to model weather, ocean and atmospheric currents, fluid flow in a pipe, and airflow around a wing, among others. The N-S equations are derived from the basics of continuity equation, momentum conservative equation, and energy conservation equation for the incompressible Newtonian fluid. The equations in the vector form with a constant viscosity μ can be expressed as

$$\rho\frac{\partial \vec{V}}{\partial t} + \rho(\vec{V}\cdot\vec{\nabla})\vec{V} = -\vec{\nabla}P + \rho\vec{g} + \mu\nabla^2\vec{V} \qquad (5.17)$$

where $\vec{\nabla}$ is the vector differential operator, $\vec{\nabla} = \vec{i}\dfrac{\partial}{\partial x} + \vec{j}\dfrac{\partial}{\partial y} + \vec{k}\dfrac{\partial}{\partial z}$, and

$$\nabla \text{ is the gradient, } \nabla = \frac{\partial}{\partial x} + \frac{\partial}{\partial y} + \frac{\partial}{\partial z}$$

TABLE 5.3

Scaling Parameters Used to Nondimensionalize the N-S Equations

Scaling Parameter	Description	Fundamental Dimension
L	Characteristic length	L
V	Characteristic velocity	Lt^{-1}
f	Characteristic frequency	t^{-1}
$P_0 - P_\infty$	Reference pressure difference	ML^{-1}/t^{-2}
g	Gravitational acceleration	Lt^{-2}

Using the scaling parameters shown in Table 5.3, defining the dimensionless variables as

$$t^* = ft \quad x^* = \frac{x}{L} \quad y^* = \frac{y}{L} \quad z^* = \frac{z}{L} \quad \vec{V}^* = \frac{\vec{V}}{V}$$

$$P^* = \frac{P - P_\infty}{P_0 - P_\infty} \quad \vec{g}^* = \frac{\vec{g}}{g} \quad \vec{\nabla}^* = L\vec{\nabla} \quad \nabla^* = L\nabla$$

and substituting the preceding dimensionless variables in Equation (5.17), the nondimensionalized N-S equation is obtained:

$$\left[\frac{fL}{V}\right]\frac{\partial \vec{V}^*}{\partial t^*} + (\vec{V}^* \cdot \vec{\nabla}^*)\vec{V}^* = -\left[\frac{P_0 - P_\infty}{\rho V^2}\right]\vec{\nabla}^* P^* + \left[\frac{gL}{V^2}\right]\vec{g}^* + \left[\frac{\mu}{\rho VL}\right]\nabla^{*2}\vec{V}^* \quad (5.18)$$

From the equation, four dimensionless numbers are obtained:

$$\text{Strouhal number: St} = \frac{fL}{V}$$

$$\text{Euler number: Eu} = \frac{P_0 - P_\infty}{\rho V^2}$$

$$\text{Froude number: Fr} = \frac{V}{\sqrt{gL}}$$

$$\text{Reynolds number: Re} = \frac{\rho VL}{\mu}$$

As a result, the nondimensionalized N-S equation with four dimensionless numbers becomes

$$[\text{St}]\frac{\partial \vec{V}^*}{\partial t^*} + (\vec{V}^* \cdot \vec{\nabla}^*)\vec{V}^* = -[\text{Eu}]\vec{\nabla}^* P^* + \left[\frac{1}{\text{Fr}^2}\right]\vec{g}^* + \left[\frac{1}{\text{Re}}\right]\nabla^{*2}\vec{V}^* \quad (5.19)$$

Alternatively, Equation (5.16) can be reorganized in the expression as

TABLE 5.4

Alternative Scaling Parameters used to Nondimensionalize the N-S Equations

Scaling Parameter	Description	Fundamental Dimension
L	Characteristic length	L
V	Characteristic velocity	Lt^{-1}
L/V	Characteristic time	t^{-1}
ρV^2	Characteristic pressure	ML^{-1}/t^{-2}
g	Gravitational acceleration	Lt^{-2}

$$\frac{\partial \vec{V}}{\partial t} + (\vec{V} \cdot \vec{\nabla})\vec{V} = -\frac{1}{\rho}\vec{\nabla}P + \vec{g} + \nu\nabla^2\vec{V} \tag{5.20}$$

where ν is the kinematic viscosity, $\nu = 1/\rho$. Using the scaling parameters shown in Table 5.4,

the dimensionless variables are defined as

$$t^* = \frac{tV}{L} \qquad x^* = \frac{x}{L} \qquad y^* = \frac{y}{L} \qquad z^* = \frac{z}{L} \qquad \vec{V}^* = \frac{\vec{V}}{V}$$

$$P^* = \frac{P}{\rho V^2} \qquad \vec{g}^* = \frac{\vec{g}}{g} \qquad \vec{\nabla}^* = L\vec{\nabla} \qquad \nabla^* = L\nabla$$

Substituting the preceding dimensionless variables in Equation (5.20), the nondimensionalized N-S equations in the vector form can be expressed as

$$\left[\frac{V^2}{L}\right]\frac{\partial \vec{V}^*}{\partial t^*} + \left[\frac{V^2}{L}\right](\vec{V}^* \cdot \vec{\nabla}^*)\vec{V}^* = -\left[\frac{V^2}{L}\right]\vec{\nabla}^* P^* + [g]\vec{g}^* + \left[\frac{\nu V}{L^2}\right]\nabla^{*2}\vec{V}^* \tag{5.21}$$

Dividing V^2/L to each term on both sides of Equation (5.21), the equation becomes

$$\frac{\partial \vec{V}^*}{\partial t^*} + (\vec{V}^* \cdot \vec{\nabla}^*)\vec{V}^* = -\vec{\nabla}^* P^* + \left[\frac{gL}{V^2}\right]\vec{g}^* + \left[\frac{\nu}{VL}\right]\nabla^{*2}\vec{V}^* \tag{5.22}$$

As a result, the nondimensionalized N-S equations with two dimensionless number, Froude number Fr and Reynolds number Re, are obtained:

$$\frac{\partial \vec{V}^*}{\partial t^*} + (\vec{V}^* \cdot \vec{\nabla}^*)\vec{V}^* = -\vec{\nabla}^* P^* + \left[\frac{1}{\text{Fr}^2}\right]\vec{g}^* + \left[\frac{1}{\text{Re}}\right]\nabla^{*2}\vec{V}^* \tag{5.23}$$

Equation (5.23) is applied in fluid mechanics particularly for the flows that the dynamic effects are dominant, that is, the high-velocity flows.

The Reynolds number is one of the most important dimensionless numbers in fluid mechanics:

$$\text{Re} = \frac{\rho V L}{\mu} \tag{5.24}$$

which is widely used to tell what the flow type is: laminar flow or turbulent flow. In Equation (5.24), L is the characteristic length. In the internal pipe flow, L is represented by a pipe diameter. In the external plate flow, L is represented by a plate length. Figure 5.5 shows the regions of flow types to the corresponding Reynolds numbers. Between laminar flow and turbulent flow, the flow is transitional. In engineering applications, the internal flow in a pipe system is typically expected in the turbulent region.

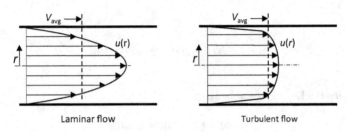

Re ≤ 2,300 Laminar flow

2,300 ≤ Re ≤ 4,000 Transition flow

Re ≥ 4,000 Turbulent flow

(a) Internal flow in a pipe

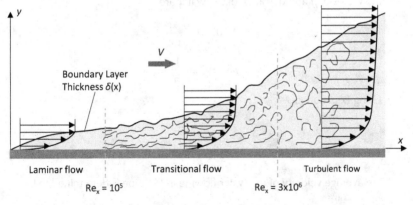

Re ≤ 10^5 Laminar flow

10^5 ≤ Re ≤ 3 × 10^6 Transition flow

Re ≥ 3 × 10^6 Turbulent flow

(b) External flow on a horizontal plate

FIGURE 5.5 Regions of flow types to the corresponding Reynolds numbers.

Exercise 5.5

Water at temperature of 24 °C flows in a pipe. The pipe diameter is 2 inches. Knowing Re = 1.7×10^4, determine (a) the flow type of the water and (b) the average velocity of the water flow in the pipe.

SOLUTION

Referring to Table 4.2 and using the unit conversion factor to convert the pipe diameter in the fundamental unit in the SI,

$$1 \text{ in} = 0.0254 \text{ m}$$

therefore,

$$2 \text{ in.} = (2 \times 0.0254 \text{ m}) = 0.051 \text{ m}$$

(a) Since the Reynolds number is

$$Re = 1.7 \times 10^5 > 4{,}000,$$

the flow type of the water in the pipe is the **turbulent flow**.

(b) At the given water temperature,

$$T = 24 \text{ °C},$$

the viscosity μ and density ρ of the water are

$$\mu = 9.107 \times 0^{-4} \frac{\text{kg}}{\text{m.s}}$$

$$\rho = 997.38 \frac{\text{kg}}{\text{m}^3}$$

Using Equation (5.24),

$$Re = \frac{\rho V D}{\mu}$$

the average velocity of the water flowing in the pipe is determined to be

$$V = \frac{\mu Re}{\rho D} = \frac{\left(9.107 \times 10^{-4} \frac{\text{kg}}{\text{m.s}}\right) 1.7 \times 10^5}{\left(997.38 \frac{\text{kg}}{\text{m}^3}\right)(0.051 \text{m})} = \frac{154.819 \frac{\text{kg}}{\text{m.s}}}{50.866 \frac{\text{kg}}{\text{m}^2}} = 3.04 \frac{\text{m}}{\text{s}}$$

Many dimensionless numbers in engineering applications can be obtained by using nondimensionalized quantity equations. Table 5.5 shows some commonly used dimensionless numbers in thermal fluid and heat transfer. Information on the dimensionless numbers named after people are listed in A.7 in the Appendix.

TABLE 5.5
Commonly Used Dimensionless Numbers in Thermal Fluid and Heat Transfer

Name	Dimensionless Expression	Ratio of Significance	Application Field
Archimedes number	$Ar = \dfrac{gL^3 \rho_s (\rho_s - \rho)}{\mu^2}$	$\dfrac{\text{Gravitational force}}{\text{Viscous force}}$	Fluid mechanics
Aspect ratio	$AR = \dfrac{L}{D}$	$\dfrac{\text{Length}}{\text{Diameter}}$	Fluid mechanics
Biot number	$Bi = \dfrac{hL}{k}$	$\dfrac{\text{Surface thermal risistance}}{\text{Internal thermal risistance}}$	Heat transfer
Capillary number	$Ca = \dfrac{\mu V}{\gamma}$	$\dfrac{\text{viscous force}}{\text{versus surface tension}}$	Fluid mechanics
Cavitation number	$Ca = \dfrac{P - P_v}{\rho V^2}$	$\dfrac{\text{Pressure} - \text{Vapor pressure}}{\text{Internal thermal risistance}}$	Fluid mechanics
Darcy friction factor	$f = \dfrac{8\tau_w}{\rho V^2}$	$\dfrac{\text{Wall friction force}}{\text{Inertial force}}$	Fluid mechanics
Drag coefficient	$C_D = \dfrac{F_D}{\frac{1}{2}\rho A V^2}$	$\dfrac{\text{Drag force}}{\text{Dynamic force}}$	Fluid mechanics
Eckert number	$Ec = \dfrac{V^2}{c_p T}$	$\dfrac{\text{Kinetic energy}}{\text{Enthalpy}}$	Heat transfer
Euler number	$Euc = \dfrac{\Delta P}{\rho V^2}$	$\dfrac{\text{Pressure difference}}{\text{Dynamic force}}$	Fluid mechanics
Fanning friction factor	$C_f = \dfrac{2\tau_w}{\rho V^2}$	$\dfrac{\text{Wall friction force}}{\text{Inertial force}}$	Fluid mechanics
Fourier number	$Fo = \dfrac{\alpha t}{L^2}$	$\dfrac{\text{diffusive rate}}{\text{storage rate}}$	Heat transfer
Froude number	$Fr = \dfrac{V}{\sqrt{gL}}$	$\dfrac{\text{Inertial force}}{\text{Gravitational force}}$	Fluid mechanics
Grashof number	$Gr = \dfrac{g\beta \Delta T L^3 \rho^2}{\mu^2}$	$\dfrac{\text{Buoyancy force}}{\text{Viscous force}}$	Heat transfer
Jakob number	$Ja = \dfrac{c_p (T - T_{sat})}{h_{fg}}$	$\dfrac{\text{Sensible energy}}{\text{Latent energy}}$	Heat transfer
Knudsen Number	$Kn = \dfrac{\lambda}{L}$	$\dfrac{\text{Mean free path length}}{\text{Characteristic length}}$	Gas dynamics

(Continued)

TABLE 5.5 (Continued)

Commonly Used Dimensionless Numbers in Thermal Fluid and Heat Transfer

Name	Dimensionless Expression	Ratio of Significance	Application Field
Laplace number	$La = \dfrac{\sigma \rho L}{\mu^2}$	Surface Tension / Momentum transport	Fluid dynamics
Lewis number	$Le = \dfrac{k}{\rho c_p D_{AB}}$	Thermal diffusion / Species diffusion	Heat transfer
Lift coefficient	$C_L = \dfrac{F_L}{\frac{1}{2}\rho A V^2}$	Lift force / Dynamic force	Fluid mechanics
Mach number	$Ma = \dfrac{V}{c}$	Flow velocity / Speed of sound	Fluid mechanics
Moment coefficient	$C_M = \dfrac{M_{LE}}{\rho a^2 C^3}$	Leading Edge moment / Wing dynamic moment	Fluid dynamics
Nusselt number	$Nu = \dfrac{Lh}{k}$	Conventional heat transfer / Conduction heat transfer	Heat transfer
Prandtl number	$Pr = \dfrac{\mu c_p}{k}$	Viscous diffusion / Thermal diffusion	Heat transfer
Pressure coefficient	$Ca = \dfrac{P - P_v}{\rho V^2}$	Static pressure difference / Dynamic pressure	Fluid mechanics
Rayleigh number	$Ra = \dfrac{g\beta\Delta T L^3 \rho^2 c_p}{k\mu}$	Buoyance force / Viscous force	Heat transfer
Reynolds number	$Re = \dfrac{\rho V L}{\mu}$	Inertial force / Viscous force	Fluid mechanics
Richardson number	$Ri = \dfrac{g\Delta\rho L^5}{\rho V^2}$	Buoyancy force / Internal energy	Fluid dynamics
Roughness ratio	$\Delta = \dfrac{\varepsilon}{L}$	Wall roughness / Characteristic length	Fluid dynamics
Specific heat ratio	$k = \dfrac{c_p}{c_v}$	Enthalpu / Internal energy	Heat transfer
Stanton number	$St = \dfrac{h}{\rho c_p V}$	Heat transfer / Thermal capacity	Heat transfer
Stokes number	$Stk = \dfrac{\rho_p D_p^2 V}{18\mu L}$	Particle relaxation time / Characteristic flow time	Fluid mechanics
Temperature ratio	$\theta = \dfrac{T_w}{T_o}$	Wall temperature / Stream temperature	Heat transfer
Ursell number	$U = \dfrac{H\lambda^2}{3}$	Surface gravigy wave / Shallow fluid layer	Fluid mechanics
Weber number	$We = \dfrac{\rho V^2 L}{\sigma_s}$	Inertial force / Surface tension force	Fluid mechanics

Exercise 5.6

Verify the Grashof number Gr used in heat transfer is a dimensionless number.

SOLUTION

Referring to Table 5.5, the Grashof number is defined as

$$Gr = \frac{g\beta\Delta T L^3 \rho^2}{\mu^2}$$

where β is the coefficient of volume expansion, $\{\beta\} = T^{-1}$;

ΔT is the temperature difference, $\{\Delta T\} = T$; and

μ is the viscosity, $\{\mu\} = ML^{-1}t^{-1}$

Fundamental dimensions of the quantities in the Gr are

g	β	ΔT	L	ρ	μ
Lt^{-2}	T^{-1}	T	L	ML^{-3}	$ML^{-3}t^{-1}$

Dimensions of the Grashof number is determined to be,

$$\{Gr\} = \frac{Lt^{-2}T^{-1}TL^3M^2L^{-6}}{M^2L^{-2}t^{-2}} = 1$$

Therefore, **the Grachof number is a dimensionless number.**

5.3   METHOD OF PI THEOROM

The method of Pi theorem appeared in 1878. The scheme of the method was proposed by Edgar Buckingham in 1914. Comparing to the method of nondimensionalized equation, the method of Pi theorem has the advantage of finding the dimensionless numbers that describe complex physical phenomena for which equations may be hard to develop. As long as the quantities involved in the phenomena are identified, the dimensionless numbers may be identified by using the Pi theorem. There are two common approaches used in the method of Pi theorem: One is called the repeating variable method, and another is called the step-by-step method.

5.3.1   THE REPEATING-VARIABLE APPROACH

The repeating-variable approach also known as the Buckingham Pi theorem named after Edgar Buckingham (1867–1940). The approach is the simplest approach to generate dimensionless numbers in engineering practice. It uses the following procedure:

- *Step 1*-Formulate a function form and count the number n of all variables in the application.

- *Step 2*-List the fundamental dimensions of all n variables.
- *Step 3*-Find the number j of the fundamental dimensions and calculate $k = n - j$, which is the expected number of dimensionless numbers.
- *Step 4*-Select j as the repeating variables being used to construct each dimensionless number except the dependent variables.
- *Step 5*-Construct k dimensionless equations and manipulate them to find k dimensionless numbers denoted as Π_1, Π_2, Π_3, and so on.
- *Step 6*-Write the final functional relationship of the dimensionless numbers in the variables.

Exercise 5.7

Knowing that a flow process described by the force F acting on an objector immersed in a stream of fluid depends on the objector length L, stream velocity V, fluid density ρ, and fluid viscosity μ, that is, $F = f(L, V, \rho, \mu)$, as shown in Figure *Exercise 5.7*, find dimensional members by using the repeating-variable approach of the Pi theorem and write the functional relationship of the dimensional members with variables.

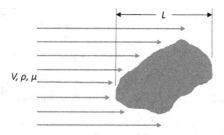

FIGURE   *Exercise 5.7.*

SOLUTION

Referring to the procedure of the repeating-variable approach,

Step 1. Count the number of variables in the application:
All relevant variables as shown in the functional form,
$F = \text{function}(L, V, \rho, \mu)$

The number of variables is

$$n = 5$$

Step 2. List the fundamental dimensions of each variable:

F	L	V	ρ	μ
$M^1 L^1 T^{-2}$	L^1	$L^1 T^{-1}$	$M^1 L^{-3}$	$M^1 L^{-1} T^{-1}$

Step 3. From Step 2, it is known that the total number of fundamental dimensions is

$$j = 3$$

They are

$$M, L, \text{ and } T$$

The expected number of dimensionless numbers, therefore, is

$$k = n - j = 5 - 3 = 2$$

Step 4. Select L, V, and ρ as three repeating variables being used to construct independent dimensionless members. F and μ are the dependent variables.

Step 5. Find k dimensionless numbers.

- The first dimensionless number Π_1 with the dependent variable F is constructed,

$$\Pi_1 = F L^{a_1} V^{b_1} \rho^{c_1}$$

In the dimensionless formula,

$$M^0 L^0 T^0 = (M^1 L^1 T^{-2})(L^1)^{a_1} (L^1 T^{-1})^{b_1}(M^1 L^{-3})^{c_1}$$

equate the exponents of each dimension,

$$\text{Mass: } M^0 = M^{1+c_1} \qquad 1 + c_1 = 0$$

$$\text{Time: } T^0 = T^{-2-b_1} \qquad -2 - b_1 = 0$$

$$\text{Length: } L^0 = L^{1+a_1+b_1-3c_1} \qquad 1 + a_1 + b_1 - 3c_1 = 0$$

Solve the preceding equations and find

$$a_1 = -2, \, b_1 = -2, \text{ and } c_1 = -1$$

Therefore, the first dimensionless number Π_1 is obtained as

$$\Pi_1 = F L^{-2} V^{-2} \rho^{-1} = \frac{F}{\rho L^2 V^2}$$

Referring to Table 5.5, the established dimensionless number Π_1 is known as the friction number (drag coefficient) C_f.

$$C_f = \frac{F}{\rho L^2 V^2}$$

- The second dimensionless member Π_2 is constructed with the dependent variable α and power -1 is selected for μ based on the experience,

$$\Pi_2 = \mu^{-1} L^{a_2} V^{b_2} \rho^{c_2}.$$

In the dimensionless formula,

$$M^0 L^0 T^0 = (M^1 L^{-1} T^{-1})^{-1} (L^1)^{a_2} (L^1 T^{-1})^{b_2} (M^1 L^{-3})^{c_2}$$

equate the exponents of each dimension,

Mass: $M^0 = M^{1+c_2}$ $c_2 - 1 = 0$

Time: $T^0 = T^{1-b_2}$ $1 - b_2 = 0$

Length: $L^0 = L^{1+a_2+b_2-3c_2}$ $1 + a_2 + b_2 - 3c_2 = 0$

solve preceding equations, and find

$$a_2 = 1,\ b_2 = 1,\ \text{and}\ c_2 = 1$$

Therefore, the second dimensionless number Π_2 is obtained as

$$\Pi_2 = L^1 V^1 \rho^1 \mu^{-1} = \frac{\rho V L}{\mu}$$

Referring to Table 5.5, the established dimensionless number Π_2 is known as the Reynolds number Re:

$$\mathbf{Re} = \frac{\rho V L}{\mu}$$

Step 6. Grouping Π_1 and Π_2, write the functional relationship of the dimensionless members in the variables as

$$\frac{F}{\rho L^2 V^2} = f\left(\frac{\rho V L}{\mu}\right)$$

Exercise 5.8

An incompressible fluid with the properties of density ρ and viscosity μ flows in a round horizontal pipe. The pipe has a length L, an inner diameter D, and an inner wall roughness height ε. The pipe is long enough that the flow is in the fully developed region as shown in Figure *Exercise 5.8*. The flow velocity of the fluid in the pipe is V. In the fully developed region, the velocity profile does not change along the pipe, but the pressure decreases linearly along the pipe due to the flow friction. If D is chosen as one of the repeating parameters, develop a nondimensional relationship of the pressure drop $\Delta P = P_1 - P_2$ with the other dimensionless numbers.

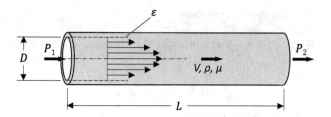

FIGURE *Exercise 5.8.*

SOLUTION

Referring to the procedure of the repeating-variable approach,

Step 1. Count the number of variables in the application:
All relevant variables as shown in the functional form,

$$\Delta P = \text{function } (L, D, V, \rho, \mu, \varepsilon)$$

The number of variables is

$$n = 7$$

Step 2. List the fundamental dimensions of each variable:

ΔP	L	D	V	μ	ε	ρ
$M^1L^{-1}T^{-2}$	L^1	L^1	L^1T^{-1}	$M^1L^{-1}T^{-1}$	L^1	M^1L^{-3}

Step 3. From Step 2, the total number of fundamental dimensions is known:

$$j = 3$$

They are

$$M, L, \text{ and } T$$

The expected number of dimensionless numbers, therefore, is

$$k = n - j = 7 - 3 = 4$$

Step 4. Any two of the parameters ε, L, and D cannot be selected as the repeating variables since their dimensions are identical. The problem statement has selected D as the repeating variable. And it is not desirable to have μ or ε appear in all dimensionless numbers. Therefore, the best choice for the repeating parameters is

$$V, D, \text{ and } \rho$$

ΔP, μ, ε, and L are the dependent variables.

Step 5. Find k dimensionless numbers.

- The first dimensionless number Π_1 is constructed with the dependent ΔP:

$$\Pi_1 = \Delta P V^{a_1} D^{b_1} \rho^{c_1} .$$

In the dimensionless formula,

$$M^0 L^0 T^0 = (M^1 L^1 T^{-2})(L^1)^{a_1} (L^1 T^{-1})^{b_1} (M^1 L^{-3})^{c_1}$$

equate the exponents of each dimension,

Mass: $M^0 = M^{1+c_1}$ $1 + c_1 = 0$

Time: $T^0 = T^{-2-a_1}$ $-2 - a_1 = 0$

Length: $L^0 = L^{-1+a_1+b_1-3c_1}$ $-1 + a_1 + b_1 - 3c_1 = 0$

Solve the preceding equations and find

$$a_1 = -2,\ b_1 = 0,\ \text{and}\ c_1 = -1$$

Therefore, the first dimensionless number Π_1 is obtained as

$$\Pi_1 = \Delta P V^{-2} D^0 \rho^{-1} = \frac{\Delta P}{\rho V^2}$$

Referring to Table 5.5, the established dimensionless number Π_1 is known as the Euler number Eu,

$$Eu = \frac{\Delta P}{\rho V^2}$$

- The second dimensionless number Π_2 is constructed with the dependent variable μ,

$$\Pi_2 = \mu V^{a_2} D^{b_2} \rho^{c_2} .$$

In the dimensionless formula,

$$M^0 L^0 T^0 = (M^1 L^{-1} T^{-1})(L^1 T^{-1})^{a_2} (L^1)^{b_2} (M^1 L^{-3})^{c_2}$$

equate the exponents of each dimension,

Mass: $M^0 = M^{1+c_2}$ $1 + c_2 = 0$

Time: $T^0 = T^{-1-a_2}$ $-1 - a_2 = 0$

Length: $L^0 = L^{-1+a_2+b_2-3c_2}$ $-1 + a_2 + b_2 - 3c_2 = 0$

Solve the preceding equations and find

$$a_2 = -1, b_2 = -1, \text{ and } c_2 = -1$$

Therefore, the second dimensionless number Π_2 is obtained as

$$\Pi_2 = \mu V^{-1} D^{-1} \rho^{-1} = \frac{\mu}{\rho V D}$$

Referring to Table 5.5, the established dimensionless number Π_3 is known as the inverse of the Reynolds number Re,

$$\text{Re} = \frac{\rho V D}{\mu}$$

- The third dimensionless number Π_3 is constructed with the dependent variable ε,

$$\Pi_3 = \varepsilon V^{a_3} D^{b_3} \rho^{c_3}$$

In the dimensionless formula,

$$M^0 L^0 T^0 = (L^1)(L T^{-1})^{a_3} (L^1)^{b_3} (M^1 L^{-3})^{c_3}$$

equate the exponents of each dimension,

Mass:	$M^0 = M^{c_3}$	$c_3 = 0$
Time:	$T^0 = T^{-a_3}$	$-a_3 = 0$
Length:	$L^0 = L^{1 + a_3 + b_3 - 3c_3}$	$1 + a_3 + b_3 - 3c_3 = 0$

Solve preceding equations and find

$$a_3 = 0, b_3 = -1, \text{ and } c_3 = 0$$

Therefore, the third dimensionless number Π_3 is obtained as

$$\Pi_3 = \varepsilon V^0 D^{-1} \rho^0 = \frac{\varepsilon}{D}$$

Referring to Table 5.5, the established dimensionless number Π_3 is known as the Roughness ratio Δ,

$$\Delta = \frac{\varepsilon}{D}$$

- The fourth dimensionless number Π_4 is constructed with the dependent L,

$$\Pi_4 = LV^{a_4}D^{b_4}\rho^{c_4}$$

In the dimensionless formula,

$$M^0L^0T^0 = (L)(LT^{-1})^{a_4}(L)^{b_4}(M^1L^{-3})^{c_4}$$

equate the exponents of each dimension,

Mass:	$M^0 = M^{c_4}$	$c_4 = 0$
Time:	$T^0 = T^{-a_4}$	$-a_4 = 0$
	$L^0 = L^{1+a_4+b_4-3c_4}$	$1 + a_4 + b_4 - 3c_4 = 0$

Solve the preceding equations and find

$$a_4 = 0, \, b_4 = -1, \text{ and } c_4 = 0$$

Therefore, the fourth dimensionless number Π_4 is obtained as

$$\Pi_4 = LV^0D^{-1}\rho^0 = \frac{L}{D}$$

Referring to Table 5.5, the established dimensionless number Π_4 is known as the aspect ratio AR:

$$AR = \frac{L}{D}$$

Step 6. Grouping Π_1, Π_2, Π_3, and Π_4, write the nondimensional relationship of the pressure drop $\Delta P = P_1 - P_2$ with the other dimensionless numbers as

$$Eu = \frac{\Delta P}{\rho V^2} = f\left(Re, \frac{\varepsilon}{D}, \frac{L}{D}\right)$$

Exercise 5.9

There is an expression to describe a R&D process as

$$H = EI^2m^{-5}G^{-2}$$

where E, I, m, and G represent the variables of energy, angular momentum, mass, and gravitational constant, respectively. Confirm that H is a dimensionless number.

SOLUTION

First, list the dimensions of all quantities:

E	I	m	G
$M^1 L^2 T^{-2}$	$M^1 L^2 T^{-1}$	M^1	$M^{-1} L^3 T^{-2}$

Then, substitute the dimensions in the expression,

$$\{H\} = (M^1 L^2 T^{-2})(M^1 L^2 T^{-1})^2 \left(M^1\right)^{-5} (M^{-1} L^3 T^{-2})^{-2}$$

After equating the exponents of each dimension,

Mass: $M^0 = M^{1 + 2 - 5 + 2}$ $1 + 2 - 5 + 2 = 0$

Time: $T^0 = T^{-2 - 2 + 4}$ $-2 - 2 + 4 = 0$

Length: $L^0 = L^{2 + 4 - 6}$ $2 + 4 - 6 = 0$

it can be seen H is dimensionless. Therefore, **H is a dimensionless number** represented by the variables of energy E, angular momentum I, mass m, and gravitational constant G.

When applying the repeating-variable approach, if the number of the fundamental dimensions j is more than or equal to the number of the variables n, then the number of the dimensional numbers becomes negative, that is, $-k = n - j (j > n)$. This is not realistic. The number of j should be adjusted to make k being positive.

Exercise 5.10

The pressure difference ΔP between the inside of a balloon P_{inside} and the outside air $P_{outside}$ can be analyzed by the repeating variables approach of the Pi theorem with other variables. The other variables are the balloon radius R and the surface tension σ_s of the balloon film are shown in Figure Exercise 5.10. Determine the relationship of ΔP in function of R and σ_s if the balloon is neutrally buoyant in the air and the gravity of the balloon is neglected.

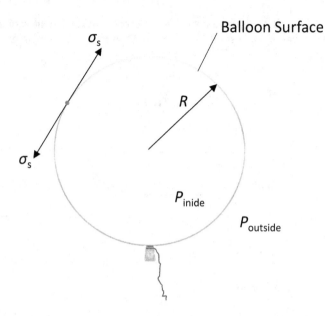

FIGURE *Exercise 5.10.*

SOLUTION

Referring to the procedure of the repeating-variable approach,

Step 1. Count the number of variables in the application:
All relevant variables can be shown in the functional form,

$$\Delta P = \text{function } (R, \sigma_s)$$

The number of variables is

$$n = 3$$

Step 2. List the fundamental dimensions of each variable:

ΔP	R	σ_s
$M^1 L^{-1} T^{-2}$	L^1	$M^1 T^{-2}$

Step 3. From Step 2, the total number of fundamental dimensions is known:

$$j = 3$$

They are

$$M, L, \text{ and } T$$

The expected number of dimensionless numbers, therefore, is

$$k = n - j = 3 - 3 = 0$$

Unfortunately, this is not logically and physically correct. To make k is positive, j may need to be reduced to two, $j = 2$:

$$k = n - j = 3 - 2 = 1$$

So, there is one dimensionless number among the variables of ΔP, R, and σ_s.

Step 4. Two repeating variables are R and σ_s, which is the only choice since ΔP is designated as the dependent variable.

Step 5. The only dependent dimensionless number Π is constructed as:

$$\Pi = \Delta P R^a \sigma_s^b$$

In the dimensionless formula,

$$M^0 L^0 T^0 = (M^1 L^{-1} T^{-2})(L^1)^a (M^1 T^{-2})^b$$

equate the exponents of each dimension,

Mass: $M^0 = M^{1+b}$ $1 + b = 0$

Time: $T^0 = T^{-2-2b}$ $-2 - 2b = 0$

Length: $L^0\} = L^{-1+a}$ $-1 + a = 0$

Solve above equations and find

$$a = 1 \text{ and } b = -1$$

Therefore, the only dimensionless number Π is obtained as

$$\Pi = \frac{\Delta P_i}{\sigma_s}$$

Step 6. Since there is only one dimensionless number, the relationship of variables ΔP, R, and σ_s in the Π must be a constant, that is,

$$\Pi = \frac{\Delta P R}{\sigma_s} = \text{constant}$$

Taking the constant to be 1, the relationship of ΔP in function of R and σ_s is expressed as

$$\Delta P = \frac{\sigma_s}{R}$$

Exercise 5.11

The voltage drop ΔE in an electrical circuit is a function of the electrical current I and the electrical resistance R. Develop the relationship of the variable ΔE in the function of I, and R in a dimensionless number and verify the relationship is the Ohm's law.

SOLUTION

Referring to the procedure of the repeating-variable approach,

Step 1. Count the number of variables in the application:
All relevant variables can be shown in the functional form,

$$\Delta E = \text{function } (I, R)$$

The number of variables is

$$n = 3$$

Step 2. List the fundamental dimensions of each variable:

ΔE	I	R
$M^1 L^2 T^{-3} I^{-1}$	I^1	$M^1 L^2 T^{-3} I^{-2}$

Step 3. From Step 2, it is known the total number of fundamental dimensions is

$$j = 4$$

They are

$$M, L, T, \text{ and } I$$

The expected number of dimensionless numbers, therefore, is

$$k = n - j = 3 - 4 = -1$$

$k = -1$ is logically and physically not correct. To make k is positive, j may need to be reduced to 3: $j = 3$.

$$k = n - j = 3 - 3 = 0$$

Unfortunately, this is not logically and physically correct either. Next, reduce j to 2, $j = 2$:

$$k = n - j = 3 - 2 = 1$$

So, there is one dimensionless number among the variables ΔE, I, and R.

Step 4. The two repeating variables of I and R are the only choice since ΔE is designated as the dependent variable.

Step 5. The only dependent dimensionless number Π is constructed as

$$\Pi = \Delta E I^a R^b$$

In the dimensionless formula,

$$M^0 L^0 T^0 = (M^1 L^2 T^{-3} I^{-1})(I^1)^a (M^1 L^2 T^{-3} I^{-2})^b$$

equate the exponents of each dimension,

Mass: $\quad\quad M^0 = M^{1+b} \quad\quad 1+b=0$

Time: $\quad\quad T^0 = T^{-3-3b} \quad\quad -3-3b=0$

Length: $\quad\quad L^0 = L^{2+2b} \quad\quad 2+2b=0$

Current: $\quad\quad I^0 = I^{-1+a-2b} \quad\quad -1+a-2b=0$

Solve the preceding equations and find

$$a = -1 \text{ and } b = -1$$

Therefore, the only dimensionless number Π is obtained as

$$\Pi = \frac{\Delta E}{IR}$$

Step 6. Since there is only one dimensionless number, the relationship of variables ΔE, I, and R in the Π must be a constant, that is,

$$\Pi = \frac{\Delta E}{IR} = \text{constant} = C$$

or

$$\Delta E = C \cdot IR$$

Taking the constant C to be 1, the relationship of ΔE in the functions of I and R is expressed as

$$\Delta E = IR$$

The preceding relationship is Ohm's law.

Exercise 5.12 (FE style)

A thermal fluid system in an application is described by four variables, length L, density ρ, velocity V, and pressure P. Determine how many dimensionless numbers can be constructed to describe the system.

(A)   1
(B)   2
(C)   3
(D)   4

SOLUTION

The correct answer is **(A)**. In total, there are four variables in the system, length L, density ρ, velocity V, and pressure P:

$$n = 4$$

Listing the fundamental dimensions of each variable:

L	ρ	V	P
L^1	M^1L^{-3}	LT^{-1}	$M^1L^{-1}T^{-2}$

The total number of fundamental dimensions is

$$j = 3$$

They are

$$M, L, \text{ and } T$$

The expected number of dimensionless numbers, therefore, is

$$k = n - j = 4 - 3 = 1$$

5.3.2 THE STEP-BY-STEP APPROACH

The step-by-step approach was developed by D. C. Ipsen in 1960. Applying the approach is straightforward and can systematically reveal all the desired dimensionless numbers at once. The procedure is simply eliminates variables involving specific dimensions in the variable relationship successively by division and multiplication, resulting in the desired dimensionless numbers. The approach does not require counting n and j as in the repeating-variable approach. For example, if the drag F acting a body in a fluid flow is known as the function of other variables, the body's characteristic length L, the body's moving velocity V, the fluid density ρ, and the fluid viscosity μ, the function relationship can be expressed as

$$F = \text{function } (L, V, \rho, \mu)$$

- *Step 1*-List all fundamental dimensions of each variable in the relationship.

F	L	V	ρ	μ
$M^1L^1T^{-2}$	L^1	L^1T^{-1}	M^1L^{-3}	$M^1L^{-1}T^{-1}$

It is known in the relationship there are three dimensions. They are

$$M, L, \text{ and } T$$

- *Step 2*-Decide to eliminate the variable with a specific dimension, for example, the mass dimension M.

- *Step 3*-Select variable ρ containing M, divide it into all other variables with the mass dimension M, and rewrite the function relationship and variable dimensions:

  Relationship: $\dfrac{F}{\rho} = \text{function}(L, \quad V, \quad \rho\!\!\!/, \quad \dfrac{\mu}{\rho})$

  Dimension: $L^4T^{-2} \quad L^1 \quad L^1T^{-1} \quad L^2T^{-1}$

  The density ρ does not exist anymore in the right side of the function relationship after Step 3.

- *Step 4*-Next, select the variable V containing T and divide it into all other variables with the time dimension T in the relationship obtained in Step 3. Observe the right side of the relationship, the velocity V is selected to be divided by appropriate powers and rewrite the function relationship and variable dimensions:

  Relationship: $\dfrac{F}{\rho V^2} = \text{function}(L, \quad V\!\!\!/, \quad \dfrac{\mu}{\rho V})$

  Dimension: $L^2 \qquad\qquad L^1 \qquad L^1$

  The velocity V does not exist anymore in the right side of the function relationship after Step 4.

- *Step 5*-Then, select the last variable containing L and divide it into all other variables with the length dimension L in the relationship obtained in Step 4. Rewrite the function relationship and variable dimension. Observe the left side of the relationship, L needs to be divided in square so that the dimensions on the right side and the left side of the function relationship are consistent.

  Relationship: $\dfrac{F}{\rho V^2 L^2} = \text{function}(L\!\!\!/, \quad \dfrac{\mu}{\rho VL})$

  Dimension: $1 \qquad\qquad\qquad 1$

  The length L does not exist anymore on the right side of the function relationship after Step 5.

- *Step 6*-Finally, the resulting relationship becomes dimensionless:

$$\frac{F}{\rho V^2 L^2} = \text{function}(\frac{\mu}{\rho VL}).$$

Two dimensionless numbers are obtained from preceding relationship. They are

$$\Pi_1 = \frac{F}{\rho V^2 L^2}$$

$$\Pi_2 = \frac{\mu}{\rho V L}$$

Referring to Table 5.5, the two established dimensionless numbers are the force coefficient C_f and the Reynold number Re, respectively, that is,

$$C_f = \frac{F}{\rho V^2 L^2}$$

and

$$\mathrm{Re} = \frac{\mu}{\rho V L}$$

Exercise 5.13

The tip deflection δ of a cantilever beam is a function of the tip load P, the beam length L, the area moment of inertia I, and the material modulus of elasticity E as shown in Figure *Exercise 5.13*. Determine the dimensionless numbers with the variables by using the step-by-step approach.

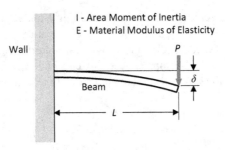

FIGURE *Exercise 5.13.*

SOLUTION

The function relationship is expressed as

$$\delta = \text{function } (P, L, I, E)$$

- *Step 1.* List all fundamental dimensions of each variable in the relationship.

δ	P	L	I	E
L^1	$M^1 L^1 T^{-2}$	L^1	L^4	$M^1 L^{-1} T^{-2}$

It is known in the relationship there are three dimensions. They are

$$M, L, \text{ and } T$$

- *Step 2.* Decide to eliminate the variable with mass dimension M.

- *Step 3.* Select variable E containing M, divide it into all other variables with the mass dimension M, and rewrite the function relationship and variable dimensions,

Relationship: $\delta = \text{function}(\dfrac{P}{E},\quad L,\quad I,\quad \cancel{E})$

Dimension: L^1 L^2 L^1 L^4

The material modulus of elasticity E does not exist anymore on the right side of the function relationship after Step 3.

- *Step 4.* Next, select the last variable containing L and divide it into all the other variables with time dimension L in the relationship obtained in Step 3. Observe the right side of the relationship, length L needs to be divided by the appropriate powers so that the dimensions on the right side and the left side of the function relationship are consistent.

Relationship: $\dfrac{\delta}{L} = \text{function}(\dfrac{P}{EL^2},\quad \cancel{L},\quad \dfrac{I}{L^4})$

Dimension: 1 1 1

Length L does not exist anymore on the right side of the function relationship after Step 4.

- *Step 5.* Finally, the resulting relationship becomes dimensionless:

$$\frac{\delta}{L} = \text{function}(\frac{P}{EL^2}, \frac{I}{L^4}).$$

Three dimensionless numbers are obtained from the preceding relationship:

$$\Pi_1 = \frac{\delta}{L}$$

$$\Pi_2 = \frac{P}{EL^2}$$

$$\Pi_3 = \frac{I}{L^4}$$

Exercise 5.14

The aerodynamic moment M_{LE} of the leading edge on a supersonic airfoil is a function of its chord length C, the angle of attack α, and the approaching velocity V as shown in Figure *Example 5.14*. Some other relevant air parameters in this application are the air density ρ, the speed of sound c, and the air-specific heat ratio k. Knowing that the effect of the air viscosity is very weak to the moment so that the variable of air viscosity is neglected in the function, determine the dimensionless numbers and the function relationship of the dimensionless numbers.

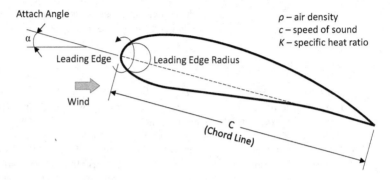

FIGURE *Exercise 5.14.*

SOLUTION

The function relationship is expressed as

$$M_{LE} = \text{function } (C, \alpha, V, \rho, c, k)$$

- *Step 1.* List all fundamental dimensions of each variable in the relationship.

M_{LE}	C	α	V	ρ	c	k
$M^1 L^2 T^{-2}$	L^1	1	$L^1 T^{-1}$	$M^1 L^{-3}$	$L^1 T^{-1}$	1

It is known that in the relationship, there are three dimensions. They are

$$M, L, \text{ and } T$$

- *Step 2.* Two of the variables in the relationship, α and k are dimensionless already. These two parameters may be presented in the derived dimensionless numbers. Decide to eliminate the variable with mass dimension M.

- *Step 3.* Select variable ρ containing M, divide it into all the other variables with mass dimension M, and rewrite the function relationship and variable dimensions,

Relationship: $\dfrac{M_{LE}}{\rho} = \text{function}(C, \quad \alpha, \quad V, \quad \not{\rho}, \quad c, \quad k)$

Dimension: $L^5 T^{-2} \qquad\qquad L^1 \quad 1 \quad L^1 T^{-1} \qquad L^1 T^{-1} \quad 1$

The density ρ does not exist anymore on the right side of the function relationship after Step 3.

- *Step 4.* Next, select variable c containing T and divide it into all the other variables with time dimension T in the relationship obtained in Step 3. The speed of sound variable a is selected to be divided by appropriate powers and rewrite the function relationship and variable dimensions:

Relationship: $\dfrac{M_{LE}}{\rho} = $ function$(C,\quad \alpha,\quad \dfrac{V}{c},\quad \varnothing,\quad k)$

Dimension: L^3 L^1 1   1   1

The speed of sound c does not exist anymore on the right side of the function relationship after Step 4.

- *Step 5.* Then, select the last variable containing L and divide it into all other variables with length dimension L in the relationship obtained in Step 4. Observe that in the left side of the function relationship, the chord variable C needs to be divided by the appropriate powers so that the dimensions on the right side and the left side of the function relationship are consistent.

Relationship: $\dfrac{M_{LE}}{\rho a^2 C^3} = $ function$(\varnothing,\quad \alpha,\quad \dfrac{V}{c},\quad k)$

Dimension: 1   1   1   1

Chord length C does not exist anymore in the right side of the function relationship after Step 5.

- *Step 6.* Finally, the resulting relationship in the variable is

$$\frac{M_{LE}}{\rho a^2 C^3} = \text{function}(\alpha, \frac{V}{c}, k).$$

Three dimensionless numbers are obtained from the preceding relationship:

$$\Pi_1 = \frac{V}{a}$$

$$\Pi_2 = \frac{C_p}{C_v}$$

$$\Pi_3 = \frac{M_{LE}}{\rho a^2 C^3}$$

Referring to Table 5.5, the three established dimensionless numbers are the Mach number Ma, the specific heat ratio k, and the moment coefficient C_M, respectively, that is,

$$\text{Ma} = \frac{V}{a},$$

$$k = \frac{c_p}{c_v}, \text{ and}$$

$$C_M = \frac{M_{LE}}{\rho a^2 C^3}$$

and the dimensionless parameter, the angle of attack α. Thus, the function relationship of the dimensionless numbers is determined to be

$$C_M = \text{function } (\alpha, \text{Ma}, k)$$

6 Equation Calculations with Units

Equation calculations are widely involved in engineering design and R&D. Numerical-value equations in engineering applications lead to confusion. For example,

$$1 + 1 = 2$$

is correct in the numerical-value equation calculation. However, the addition of $1 + 1$ may not be equal to 2 in the quantity equation calculation since it may have

$$1 \text{ in.} + 1 \text{ ft} = 13 \text{ in., or } 13/12 \text{ ft}$$

Similarly, the multiplication of

$$12 \times 10 = 120$$

is correct in the numerical-value equation calculation. But 12×10 may not be equal to 120 in a quantity equation calculation since it may have

$$12 \text{ in.} \times 10 \text{ ft} = 10 \text{ ft}^2 \text{ or } 1{,}440 \text{ in.}^2$$

While a quantity equation describes a relation among different physical quantities, in the equation, the quantity is expressed in two portions: the numerical-value portion, which simply is the magnitude of the quantity, and the unit portion (see Section 1.1 in Chapter 1). The unit is an important part of equation calculations. When performing an equation calculation, showing the units throughout the entire calculation and ensure that the units are consistent throughout the calculation is advisable. The units shown in calculations are also helpful for better understanding and conveniently checking if the calculation errors occur. By showing units at the end of a calculation or in the final result, the tracking of errors caused by unit conversions the calculation may be difficult.

6.1 UNIT CONSISTENCY

Keeping unit consistency in equation calculations is important. The advantages of showing units in equation calculations are obvious. Frequently, a wrong result is caused by calculation errors from unit inconsistency. For example, a manometer is attached to a pressurized air tank under the atmospheric pressure of 100 kPa. In the

DOI: 10.1201/9781003508977-6

tank, the air absolute pressure is 76 kPa. Knowing that the water density ρ_w is 1,000 kg/m³, and the mercury specific gravity SG_{Hg} is 13.55 as shown in Figure 6.1, it needs to find the unknown SG_{liquid} in the inclined tube.

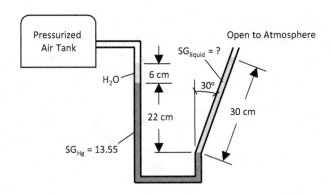

FIGURE 6.1 Pressure measurement.

Solution (without showing units in equation calculation and incorrect result)

Using the pressure balance along the tube,

$$P_{air} + P_w + P_{Hg} - P_{liquid} - P_{atm} = 0 \qquad (6.1)$$

and knowing that

$$P = \rho g h,$$

$$\rho = SG \rho_w, \text{ and}$$

$$\rho_w = 1,000 \text{ kg/m}^3$$

Equation (6.1) becomes

$$P_{air} + \rho_w g h_w + \rho_{hg} g h_{Hg} - \rho_{liquid} g h_{liquid} - P_{atm} = 0 \qquad (6.2)$$

Subsuming values in Equation (6.2),

$$76 + 9.81 \rho_w \times 6 \times 10^{-2} - 9.81 \rho_{Hg} \times 22 \times 10^{-2} - 9.81 \times \rho_{liquid}$$
$$\times 30 \times 10^{-2} \times \cos 30^\circ - 100 = 0$$
$$76 + 0.589 \rho_w + 2.158 \rho_{Hg} - 2.6 \rho_{liquid} - 100 = 0 \qquad (6.3)$$
$$565 = 2.6 \rho_{liquid} - 2.158 \rho_{Hg} \qquad (6.4)$$

Because

$$\rho_{Hg} = \rho_w SG_{hg}$$
$$\rho_{hg} = 1,000 \times 13.55 = 13,550$$

Equation (6.4) becomes

$$565 = 2.6\rho_{liqud} - 2.158 \times 13,550$$
$$P_{liquid} = \frac{2.158 \times 13,550 + 565}{2.6} = \frac{29,805.9}{2.6} = 11,463.8$$

Finally,

$$SG_{oil} = \frac{\rho_{liquid}}{\rho_w} = \frac{11,463.8}{1,000} = 11.464. (\text{This is a wrong result})$$

Analysis

The error in the preceding calculation is from unit inconsistent. In Equation (6.3),

$$76 + 0.589\rho_w + 2.158\rho_{Hg} - 2.6\rho_{liquid} - 100 = 0$$

the first term, second term, and last term cannot be added because the units of 76 and 100 are kPa, while the unit of $0.589\rho_w$ is Pa. After correctly manipulating Equation (6.4), the result, in kPa, should be

$$-24 + 0.589 + 2.158 \times 13.55 = \left(\frac{2.6}{1,000}\right) SG_{liquid}\rho_w$$

or, in Pa,

$$-24.000 + 0.589\rho_w + 2.158 \times 13.55\rho_w = 2.6\, SG_{liquid}\rho_w$$

Consequently, **the correct SG_{liquid} is 2.3.**

Solution (showing units in equation calculations and having the correct result)

Using the pressure balance along the tube,

$$P_{air} + P_w + P_{Hg} - P_{liquid} - P_{atm} = 0$$

and knowing that

$$P = \rho g h,$$
$$\rho = SG\rho_w,$$
$$\rho_w = 1,000 \text{ kg/m}^3$$

then

$$P_{air} + \rho_w g h_w + \rho_{hg} g h_{Hg} - \rho_{liquid} g h_{liquid} - P_{atm} = 0$$

Substituting values with units into preceding equation,

$$76(kPa) + 1,000(kg/m^3)9.81(m/s^2)[6\times10^{-2}(m) + SG_{hg}22\times10^{-2}(m)$$
$$- SG_{liquid}30\times10^{-2}\cos 30^\circ (m)] - 100(kPa) = 0$$

and after manipulations, the equation becomes

$$-24(kPa) + 9,810(kg/m^2.s^2)(6\times10^{-2} + 13.55\times22\times10^{-2} - 0.26SG_{liquid})(m) = 0$$
$$-24(kPa) + 9,810(kg/m.s^2)(3.041 - 0.26SG_{liquid}) = 0$$

Knowing that

$$1\ kPa = 1,000\ Pa$$
$$1\ kg/m.s^2 = 1\ Pa$$

so

$$-24,000(Pa) + 9,810(Pa)(3.041 - 0.26SG_{liquid}) = 0$$
$$0.26SG_{liquid} = 3.041 - \frac{24,000(Pa)}{9,810(Pa)} = 3.041 - 2.447 = 0.594$$

Finally, the unknown SG_{liquid} is obtained:

$$SG_{liquid} = \frac{0.594}{0.26} = 2.3$$

The preceding example clearly shows including units in equation calculations is a good practice in engineering applications, which definitely helps with keeping the unit consistent and tracking errors in calculations.

Exercise 6.1

A centrifugal compressor compresses 60 lbm/s airflow at a rotation speed of 11,000 rpm, and a rotor diameter D at the compressor exit is 2 ft as shown in the Figure *Exercise 6.1*. Determine (a) the torque τ on the rotor blade at the exit and (b) the horsepower P required to drive the compressor.

SOLUTION

(a) Since the rotation speed $\dot{n} = 11,000$ rpm and the rotor radius $R = 0.5\ D = 1$ ft, the tangential velocity of the root blade at exit is

$$V_t = 2\pi R\left(\frac{\dot{n}}{60}\right)$$
$$= 2\pi \times 1\left(\frac{11,000}{60}\right) = 1,151.92\frac{ft}{s}.$$

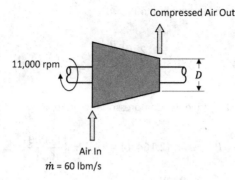

Compressed Air Out

11,000 rpm

D

Air In

$\dot{m} = 60$ lbm/s

FIGURE *Exercise 6.1.*

The torque on the rotor blade is obtained:

$$\tau = \dot{m}RV_t$$

$$= \frac{60}{32.2}(1 \times 1,151.92) = \mathbf{2,164.44 \ ft \cdot lbf}$$

(b) The angular velocity is determined to be

$$\omega = \frac{V_t}{R} = \frac{2\pi R\left(\dfrac{\dot{n}}{60}\right)}{R} = 2\pi\left(\frac{\dot{n}}{60}\right)$$

The power required to drive the compressor, therefore, is

$$P = \tau\omega = \frac{2,164.44(2\pi)\left(\dfrac{11,000}{60}\right)}{550} = \mathbf{4,533.19 \ hp}$$

The previous calculation, however, may cause difficult for understanding and checking if the conversion factors of 1 min = 60 s, 1 hp = 550 lbf.ft/s, and g_c = 32.2 lbm.ft/lbf·s² can't be acknowleged for using to keep the unit consistency in the calculation. Obviously, presenting all units in the entire calculation is preferable for understanding and convenience as shown:

$$V_t = 2\pi R\left(\frac{\dot{n}}{60}\right)$$

$$= 2\pi \times (1 \ ft)\left(\frac{11,000}{60 \ s}\right) = 1,151.92\frac{ft}{s}$$

The torque on the rotor blade is obtained:

$$\tau = \dot{m}RV_t$$

$$= \frac{60\dfrac{lbm}{s}}{32.2\dfrac{lbm \cdot ft}{lbf \cdot s^2}}\left(1 \ ft \times 1,151.92\frac{ft}{s}\right) = 2,164.44 \ ft \cdot lbf$$

The angular velocity is determined to be

$$\omega = \frac{V_t}{R} = \frac{2\pi R\left(\dfrac{\dot{n}}{60}\right)}{R} = 2\pi\left(\frac{\dot{n}}{60}\right)$$

The power required to drive the compressor, therefore, is

$$P = \tau\omega = (2{,}164.44 \text{ ft·lbf})(2\pi)\left(\frac{11{,}000}{60 \text{ s}}\right)$$

$$= 2{,}493{,}255.89 \frac{\text{ft·lbf}}{\text{s}}$$

In horsepower, it is

$$p = \left(2{,}493{,}255.89 \frac{\text{ft·lbf}}{\text{s}}\right)\left(\frac{1\,\text{hp}}{550\dfrac{\text{ft·lbf}}{\text{s}}}\right) = 4{,}533.19 \text{ hp}$$

Exercise 6.2

The 32.2 lbm carriage A moves horizontally in a guide at a speed of 5 ft/s. The carriage carries two assemblies with balls and light rods that rotate about a shaft at a point O on both sides of the carriage as shown in Figure *Exercise 6.2*. Knowing that each of the four balls weighs 3.22 lbf, the assembly on the front side rotates counterclockwise at a speed of 60 rev/min, and the assembly on the back side rotates clockwise at a speed of 80 rev/min, determine the kinetic energy T of the entire system if any friction on the guide and from rotation are neglected.

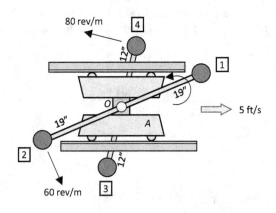

FIGURE *Exercise 6.2.*

SOLUTION

The velocities of the balls with respect to the O point are expressed as

$$V_{rel} = r\dot{\theta}$$

where V_{rel} is the relative velocity, r is the distance from the ball to the fixed point, and $\dot{\theta}$ is the angular velocity, respectively.
Then,

$$\left(V_{rel}\right)_{1,2} = \frac{19}{12}\frac{60(2\pi)}{60} = 9.95\,\text{ft}/\text{s}$$

$$\left(V_{rel}\right)_{3,4} = \frac{12}{12}\frac{80(2\pi)}{60} = 8.38\,\text{ft}/\text{s}$$

The translational part of the kinetic energy of the system, therefore, is

$$\frac{1}{2}m\bar{V}^2 = \frac{1}{2}\left(\frac{32.2}{32.2} + 4\frac{3.22}{32.2}\right)(5^2) = 17.5\,\text{ft·lbf}$$

The rotational part of the kinetic energy of the system, depending on the squares of the relative velocities, is determined to be

$$\sum \frac{1}{2}m_i|\dot{\rho}_i|^2 = 2\left[\frac{1}{2}\frac{3.22}{32.2}(9.95)^2\right]_{(1,2)} + 2\left[\frac{1}{2}\frac{3.22}{32.2}(8.38)^2\right]_{(3,4)}$$

$$= 9.90 + 7.02 = 16.92\,\text{ft·lbf}$$

The total kinetic energy of the system is obtained:

$$T = \frac{1}{2}m\bar{V}^2 + \sum \frac{1}{2}m_i|\dot{\rho}_i|^2 = 17.5 + 16.92 = \mathbf{34.42\ ft·lbf}$$

In the preceding calculation, it may lead to confusion about why the numerical values 12, 60, and 32.2 are used in the equations for finding $\left(V_{rel}\right)_{1,2}$, $\left(V_{rel}\right)_{3,4}$, $\frac{1}{2}m\bar{V}^2$, and $\sum \frac{1}{2}m_i|\dot{\rho}_i|^2$. These values actually are used for the unit conversion to keep unit consistency in the calculations. Obviously, showing the units in the calculation is better for understanding and convenience, such as

$$\left(V_{rel}\right)1,2 = 19\text{in}\frac{1\text{ft}}{12\text{in}}\frac{\left(60\frac{\text{rev}}{\text{min}}\right)\left(\frac{2\pi}{\text{rev}}\right)}{60\frac{\text{s}}{\text{min}}} = 9.95\,\text{ft}/\text{s}$$

$$\left(V_{rel}\right)3,4 = 12\text{in}\frac{1\text{ft}}{12\text{in}}\frac{\left(80\frac{\text{rev}}{\text{min}}\right)\left(\frac{2\pi}{\text{rev}}\right)}{60\frac{\text{s}}{\text{min}}}\frac{12}{12} = 8.38\,\text{ft}/\text{s}$$

$$\frac{1}{2}m\bar{V}^2 = \frac{1}{2}\left(\frac{32.2\text{lbm}}{32.2\frac{\text{lbm·ft}}{\text{lbf·s}^2}} + 4\frac{3.22\text{lbm}}{32.2\frac{\text{lbm·ft}}{\text{lbf·s}^2}}\right)\left(5\frac{\text{ft}}{\text{s}}\right)^2 = 17.5\,\text{ft·lbf}$$

$$\sum \frac{1}{2}m_i\left|\dot{\rho}_i\right|^2 = 2\left[\frac{1}{2}\frac{3.22\text{lbm}}{32.2\dfrac{\text{lbm}\cdot\text{ft}}{\text{lbf}\cdot\text{s}^2}}\left(9.95\frac{\text{ft}}{\text{s}}\right)^2\right]_{(1,2)} + 2\left[\frac{1}{2}\frac{3.22\text{lbm}}{32.2\dfrac{\text{lbm}\cdot\text{ft}}{\text{lbf}\cdot\text{s}^2}}\left(8.38\frac{\text{ft}}{\text{s}}\right)\right)^2\right]_{(3,4)}$$

$$= 9.90\,\text{ft·lbf} + 7.02\,\text{ft·lb}$$
$$= 16.92\,\text{ft·lbf}$$

The total kinetic energy of the system, therefore, is

$$T = \frac{1}{2}m\bar{V}^2 + \sum\frac{1}{2}m_i\left|\dot{\rho}_i\right|^2 = 17.5\,\text{ft·lbf} + 16.92\,\text{ft·lbf}$$
$$= 34.42\,\text{ft·lbf}$$

Exercise 6.3

Calculating the geometric parameters of the wing for the space shuttle, the complex shape of the actual wing is replaced by a swept, trapezoidal wing (reference wing) as shown in Figure *Exercise 6.3*. In the reference wing area, the tip chord ct is 16.53 ft, the root chord cr is 60.28 ft, and the span S is 80.34 ft. Determine (a) the wing area A, (b) the aspect ratio AR, and (c) the taper ratio δ.

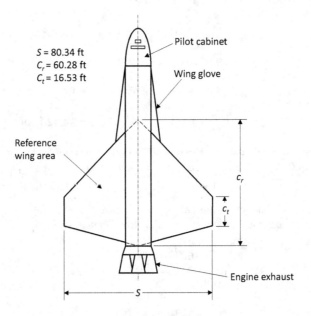

FIGURE *Exercise 6.3.*

SOLUTION

Per the given data of the reference wing, the following calculations are conducted.

(a) The area of the trapezoidal reference wing is

$$A = 2\left(\frac{C_t + C_r}{2}\right)\frac{s}{2}$$

$$= 2\left(\frac{16.53\,\text{ft} + 60.28\,\text{ft}}{2}\right)\frac{80.34\,\text{ft}}{2} = 3{,}085.46\,\text{ft}^2$$

(b) The aspect ratio AR for this swept of the trapezoidal wing is

$$AR = \frac{s^2}{A}$$

$$= \frac{(80.34\,\text{ft})^2}{3{,}085.46\,\text{ft}^2} = 2.12$$

(c) The taper ratio is

$$\delta = \frac{C_t}{C_r}$$

$$= \frac{16.53\,\text{ft}}{60.28\,\text{ft}} = 0.27$$

Any quantity equations must have the same dimensions and units on the left and right sides of the calculation. To keep unit consistency, unit conversions may inevitably be involved. In general, there are two approaches to conduct unit conversions:

- Converting units to the standard units or the designated units prior to equation calculations.
- Converting units for unit consistency during equation calculations.

6.1.1 Using Standard Units Prior to Calculations

This approach is to use unit conversion factors (see Section 4.1 in Chapter 4) to convert units to the standard units (fundamental and derived) prior to equation calculations. For example, convert units to the standard units,

- force (N), mass (kg), length (m), and pressure (kPa) in the SI and
- force (lbf), mass (lbm), length (ft), and pressure (psi) in the English system,

or convert units to the designated units of the problem being solved.

Exercise 6.4

Referring back to Figure 6.1, the manometer is attached to a pressurized air tank as shown under the atmospheric pressure of 100 kPa. In the tank, the air absolute pressure is 76 kPa. Knowing that water density ρ_w is 1,000 kg/m³, and mercury specific gravity SG_{Hg} is 13.55, find the unknown SG_{liquid} in the inclined tube by using the units in the English system.

SOLUTION

Referring to the Table 4.5 by using the unit conversion factors to convert all units to the standard units in the English system,

$$1\,kPa = 0.14696\,psi$$

$$1\frac{kg}{m^3} = 0.062428\frac{lbm}{ft^3}$$

$$1\,cm = 0.032808\,ft$$

therefore,

$$76\,kPa = (76 \times 0.14696)\,psi = 11.172\,psi$$

$$100\,kPa = (100 \times 0.14696)\,psi = 14.696\,psi$$

$$1{,}000\frac{kg}{m^3} = (1{,}000 \times 0.062428)\frac{lbm}{ft^3} = 62.428\frac{lbm}{ft^3}$$

$$6\,cm = (6 \times 0.032808)\,ft = 0.197\,ft$$

$$22\,cm = (22 \times 0.032808)\,ft = 0.722\,ft$$

$$30\,cm = (30 \times 0.032808)\,ft = 0.984\,ft$$

Using Equation (6.2) in Section 6.1,

$$P_{air} + \rho_w g h_w + \rho_{hg} g h_{Hg} - \rho_{liquid} g h_{liquid} - P_{atm} = 0$$

$$11.172\ psi + \left(62.428\frac{lbm}{ft^3}\right)\left(32.174\frac{ft}{s^2}\right)(0.197\ ft)$$

$$+\ 13.55 \times \left(62.428\frac{lbm}{ft^3}\right)\left(32.174\frac{ft}{s^2}\right)(0.722\ ft)$$

$$-\ SG_{liquid}\left(62.428\frac{lbm}{ft^3}\right)\left(32.174\frac{ft}{s^2}\right)(0.984\ ft)\cos 30°$$

$$-14.696\ psi = 0$$

$$11.172\ psi + \left(62.428\frac{lbm}{ft^3}\right)\left(32.174\frac{ft}{s^2}\right)(0.197\ ft)$$

$$+\ (13.55 \times 0.722\,ft) - SG_{liquid}\,(0.984\,ft)\cos 30°$$

$$-14.696\,psi = 0$$

$$-3.5\ psi + 2{,}008.6\frac{lbm}{ft^2 \cdot s^2}\left[9.98\ ft - (0.85\ ft)SG_{liquid}\right] = 0 \tag{E6.4-1}$$

$$-3.5\ psi + 20{,}045.83\frac{lbm}{ft \cdot s^2} - \left(1{,}707.3\frac{lbm}{ft \cdot s^2}\right)SG_{liquid} = 0$$

and the unit conversion factors referred to Table 4.1,

$$1\text{psi} = 144\frac{\text{lbf}}{\text{ft}^2}$$

$$1\text{lbf} = 32.174\frac{\text{lbm·ft}}{\text{s}^2}$$

therefore,

$$3.5\text{ psi} = (3.5 \times 144 \times 32.174)\frac{\text{lbm}}{\text{ft·s}^2} = 16{,}15.7\frac{\text{lbm}}{\text{ft·s}^2}$$

Substituting the preceding expression in Equation (E6.4–1),

$$-16{,}215.7\frac{\text{lbm}}{\text{ft·s}^2} + 20{,}045.83\frac{\text{lbm}}{\text{ft·s}^2} - \left(1{,}707.3\frac{\text{lbm}}{\text{ft·s}^2}\right)SG_{\text{liquid}} = 0$$

Finally, the specific gravity of the liquid is obtained to be

$$SG_{\text{liquid}} = \frac{3{,}830.13\dfrac{\text{lbm}}{\text{ft·s}^2}}{1{,}707.3\dfrac{\text{lbm}}{\text{ft·s}^2}} \doteq \mathbf{2.3}$$

Exercise 6.5 (FE style)

The equation to calculate the length change Δl of a pipe by thermal expansion is described by

$$\Delta l = l \times \alpha \times \Delta T$$

where l is the existing pipe length, α is the thermal expansion coefficient, and ΔT is the temperature difference of T_{max} and T_{min} of the pipe in operation. If the pipe in the operation has the following specifications,

Pipe material: Steel,

Pipe length: 30 m,

$T_{max} = 353$ K, and

$T_{min} = 288$ K

and α is selected from the following table, the length change ΔL of the pipe by thermal expansion in meters will be

Material	Expansion Coefficient α (mm/m.°C)
Steel	0.0120
Stainless steel	0.0166
Copper	0.0168
Aluminum	0.0232

(A) 0.0205 m.
(B) 0.0218 m.
(C) 0.0227 m.
(D) 0.0234 m.

SOLUTION

The correct answer is **(D)**. Using the unit conversion factors,

$$T(°C) = T(°K) - 273$$

$$1 \text{ mm} = 0.001 \text{ m}$$

The temperatures of T_{max} and T_{min} are

$$T_{max} = (353 - 273) \text{ °C} = 80 \text{ °C}$$

$$T_{min} = (288 - 273) \text{ °C} = 15 \text{ °C}$$

The expansion coefficient α found in the table is

$$0.0120 \frac{\text{mm}}{\text{m.°C}} = 0.0120 \times 0.001 \frac{\text{m}}{\text{m.°C}} = 1.2 \times 10^{-5} \frac{\text{m}}{\text{m.°C}}$$

The length change of the pipe is determined to be

$$\Delta l = (30 \text{ m})\left(1.2 \times 10^{-5} \frac{\text{m}}{\text{m.°C}}\right)(80 - 15)° \text{ C}$$

$$= \mathbf{0.0234\,m}$$

Alternatively, using the unit conversion factor to convert millimeters to meters,

$$1 \text{ m} = 1,000 \text{ mm}$$

then,

$$30 \text{ m} = 30,000 \text{ mm}$$

$$0.012 \frac{\text{mm}}{\text{m.°C}} = 0.0120 \times 10^{-3} \frac{\text{mm}}{\text{mm.°C}} = 1.2 \times 10^{-5} \frac{\text{mm}}{\text{mm.°C}}$$

the length change of the pipe is

$$\Delta l = (30,000 \text{ mm})\left(1.2 \times 10^{-5} \frac{\text{mm}}{\text{mm.°C}}\right)(80 - 15)° \text{ C}$$

$$= 23.4\text{mm}$$

Since

$$1 \text{ mm} = 0.001 \text{ m}$$

the length change of the pipe in meters is

$$\Delta l = 23.4 \text{ mm} = 23.4 \times 0.001 \text{ m}$$
$$= 0.0234 \text{ m}.$$

6.1.2 CONVERTING UNITS DURING CALCULATIONS

Using unity conversion ratios (see Section 4.2 in Chapter 4) in the equation calculations is an efficient way to convert units. For example, referring to *Exercise 6.5* and using the unity conversion ratio,

$$\frac{1 \text{m}}{1,000 \text{mm}}$$

the length change of the pipe can be determined conveniently to be

$$\Delta l = (30 \text{m}) \left(0.0120 \frac{\text{mm}}{\text{m.}^\circ\text{C}} \times \frac{1 \text{m}}{1,000 \text{mm}} \right) (80 - 15)^\circ \text{C}$$
$$= 0.0234 \text{m}$$

Exercise 6.6

An airplane with a weight $W = 420,000$ lbf is cruising at a speed $V = 220$ kn at an altitude $h = 22,000$ ft as shown in Figure *Exercise 6.6*. Determine the total energy the airplane possesses when neglecting any frictions and other energy losses.

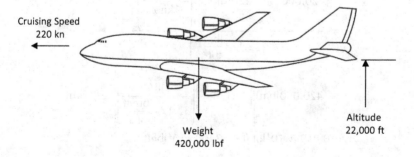

Cruising Speed
220 kn

Weight
420,000 lbf

Altitude
22,000 ft

FIGURE *Exercise 6.6.*

SOLUTION

The problem uses the units of the English system. Referring to Table 4.1 and using unity conversion ratios,

$$\frac{1.6878\frac{ft}{s}}{1\,knot}$$

$$\frac{1\,lbf}{32.174\frac{lbm\cdot ft}{s^2}}$$

and the equation of mass,

$$m = \frac{W}{g}$$

the airplane mass is obtained to be

$$m = \frac{420{,}000\,lbf}{32.174\frac{ft}{s^2}}\left(\frac{32.174lbm\cdot\frac{ft}{s^2}}{1lbf}\right) = 420{,}000\,lbm$$

The equation of the total energy E is

$$E = \frac{mV^2}{2g_c} + mgh$$

Therefore,

$$E = \frac{(420{,}000\,lbm)(220\,knots)^2\left(\frac{1.6878\frac{ft}{s}}{1\,knot}\right)^2}{2\left(32.174\frac{lbm\cdot ft}{lbf\cdot s^2}\right)}$$

$$+(420{,}000lbm)\left[\left(32.174\frac{ft}{s^2}\right)\left(\frac{1lbf}{32.174\frac{lbm\cdot ft}{lbf\cdot s^2}}\right)\right](22{,}000\,ft)$$

$$= 0.8999\times10^9\,lbf\cdot ft + 9.2400\times10^9\,lbf\cdot ft$$

$$= 10.1399\times10^9\,lbf\cdot ft$$

Exercise 6.7

An engineer using the units in the English system to conduct an energy audit for a building with brick walls. The wall of the building is 20 cm as shown in Figure *Exercise 6.7*. The temperatures on the outside wall are 33 °C and 22 °C, respectively. Auditing to convert the unit of the thermal conductivity k of the brick from the SI to the English system is necessary. Knowing that the thermal conductivity of the brick is 0.76 W/m °C from the handbook, determine (a) k in the unit of Btu/h. ft °F and (b) the rate of heat transfer in the unit of Btu/h through the wall area per square foot.

SOLUTION

Referring to Table 4.5 and using the unity conversion ratios,

$$\frac{3.412\,\dfrac{Btu}{h}}{1\,w},$$

$$\frac{3.2808\,ft}{1\,m},$$

$$\frac{1.8°F}{1°C}, \text{ and}$$

$$\frac{0.032808\,ft}{1\,cm}$$

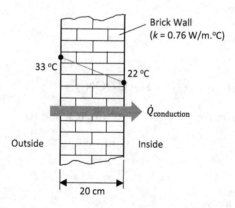

FIGURE *Exercise 6.7.*

(a) The thermal conductivity of the brick is decided to be

$$k = 0.76 \frac{W}{m.°C} = 0.76 \frac{w\left(\dfrac{3.412\dfrac{Btu}{h}}{1\,w}\right)}{m\left(\dfrac{3.2808\,ft}{1\,m}\right).°C\left(\dfrac{1.8\,°F}{1\,°C}\right)}.$$

$$= (0.76 \times 0.5777)\frac{Btu}{h.ft.°F} = \mathbf{0.4391}\frac{\mathbf{Btu}}{\mathbf{h.ft.°F}}$$

(b) The rate of heat conduction is determined as

$$\dot{Q}_{conduction} = KA\frac{T_o - T_i}{L}$$

$$= \left(0.4391\frac{Btu}{h.ft.\,°F}\right)(1ft^2)\frac{\left[(33\,°C - 22\,°C)\left(\dfrac{1.8\,°F}{1\,°C}\right)\right]}{(2\,cm)\dfrac{0.03208\,ft}{1\,cm}}.$$

$$= \mathbf{135.51 Btu\,/\,h}$$

Exercise 6.8 (FE style)

An air stream has the pressure $P = 2,000$ lbf/ft^2 and temperature $T = 40$ °F. If knowing the gas constant of air $R = 53.34$ ft.lbf/lbm.R, select all correct answers for the air density shown.

(A) $0.075\,lbm\,/\,ft^3$

(B) $1.2014\dfrac{kg}{m^3}$

(C) $4.34 \times 10^{-5}\dfrac{lbm}{in^3}$

(D) $2.233 \times 10^{-3}\dfrac{slugs}{ft^3}$

SOLUTION

The correct answers are all of them, **(A)**, **(B)**, **(C)**, and **(D)**. Using the ideal gas equation of state

$$\rho = \frac{P}{RT}$$

the density of the air is

$$\rho = \frac{2,000\dfrac{lbf}{ft^2}}{\left(53.34\dfrac{ft \cdot lbf}{lbm \cdot R}\right)(40 + 460)R} = 0.075\frac{lbm}{ft^3}$$

Referring to Table 4.5 and using unity conversion ratios,

$$\frac{16.0185\,\frac{kg}{m^3}}{1\frac{lbm}{ft^3}},$$

$$\frac{1{,}728\,in^3}{1\,ft^3}$$

and

$$\frac{0.031\,slugs}{1\,lbm}$$

the density of the air is determined to be

$$\rho = \left(0.0750\,\frac{lbm}{ft^3}\right)\left(\frac{16.0185\,\frac{kg}{m^3}}{1\frac{lbm}{ft^3}}\right) = 1.2014\,\frac{kg}{m^3},$$

$$\rho = \left(0.0750\,\frac{lbm}{ft^3}\right)\left(\frac{lbm}{ft^3\left(\frac{1{,}728\,in^3}{1\,ft^3}\right)}\right) = 4.34\times10^{-5}\,\frac{lbm}{in^3}$$

and

$$\rho = \left(0.0750\,\frac{lbm}{ft^3}\right)\left(\frac{\frac{0.031\,slugs}{1\,lbm}}{ft^3}\right) = 2.33\times10^{-3}\,\frac{slugs}{ft^3}$$

6.2 SIGNIFICANT DIGITS

Unit conversion involves the calculations of numerical values. The numerical values can be integers in hundred thousands or decimals in hundred thousandths. Keeping long digits of the numerical value is not necessary, and they should be rounded to a certain place without losing the accuracy of value. The degree of accuracy of the value depends on how many digits are kept. After a certain place, the digits may not be significant. Significant digits, also called significant figures, are the number of digits to contribute the accuracy and precision of measurement. The concept of the significant digits is used to decide how many digits in the numerical value need to be reserved for an accurate measurement. In engineering practice, some basic rules are used to decide how many digits should be preserved as significant in a numerical value or a number.

Significant Digits

- All nonzero digits are significant.
- Any zeroes between the nonzero digits are significant.
- Zeroes to the right of a decimal point called the trailing zeroes are significant.

Figure 6.2 shows an illustration of the significant digits by using the number 510.0300100.

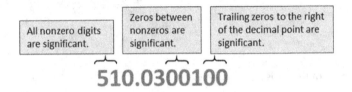

FIGURE 6.2 Significant digits.

Nonsignificant Digits

The only digits that are not significant are zeros that are acting only as placeholders in a number:

- Trailing zeros in a whole number without a decimal shown are not significant.
- Leading zeros after the first nonzero digit on the right of a decimal point are not significant. They only represent the position of the decimal point.

Figure 6.3 shows an illustration of the nonsignificant digits in the numbers 5,100 and 0.03001.

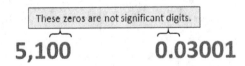

FIGURE 6.3 Nonsignificant digits.

Exercise 6.9

Estimate how many significant figures are in the following physical quantities.

1. 10.007,5 lbm
2. 10.007,500 lbm
3. 0.007,5 lbm
4. 5,000 lbf
5. 5,000.00 kg

6. 1.234 lbm
7. 0.012 kJ
8. 0.250 m
9. 3.07 m/s
10. 0.001 °C

SOLUTION

1. There are six significant digits. All zeros between the nonzero digits are significant.
2. There are eight significant digits. The trailing zeros to the right of the decimal point are significant.
3. There are two significant digits. Leading zeros to the right of the decimal point only indicate the position of the decimal point.
4. There is only one significant digit. Trailing zeros in the whole number without a decimal shown are not significant
5. There are six significant digits. The trailing zeros to the right of a decimal point are significant.
6. There are four significant figures. All digits are nonzero.
7. There are two significant figures. Leading zeros to the right of the decimal point, which merely indicate the position of the decimal point.
8. There are three significant figures. The trailing zero to the right of a decimal point is significant.
9. There are three significant figures. All nonzero digits are significant and any zeroes between the nonzero digits are significant.
10. There is only one significant figure. Leading zeros after the first nonzero digit on the right of a decimal point are not significant. They only represent the position of the decimal point.

Exercise 6.10 (FE style)

Determine which physical quantities shown have only one significant digit.

(A) 0.1 ft
(B) 0.01 kg
(C) 0.001 kN
(D) 0.0001 kW

SOLUTION

The correct answers are **(A)**, **(B)**, **(C)**, and **(D)**. All of them have one significant digit since the leading zeros to the right of the decimal point were not counted as significant digits. These zeros merely indicate the position of the decimal point.

Mathematical Operations

In equation calculations, the rule is that the accuracy of a calculated result is limited by the least accurate measurement involved in the calculation. In general, the least number of significant digits in the given numbers determines the number of significant digits in mathematical operations. The steps of determining the number of significant digits follow:

1. Convert all terms in the operation to the same units.
2. Identify the significant digits of each number in the operation to the right of the decimal point.
3. Find the least number of significant digits to the right of the decimal point of all numbers.
4. Carry out the operation of all numbers in normal manners.
5. Complete the operation and round off the number to the least significant digits (see Section 6.3 in Chapter 6) for rounding off.
6. The number of significant digits determined in operation is the least number of significant digits of numbers.

Exercise 6.11 (addition and subtraction)

Find the final numerical value and determine the number of significant digits for the following calculations.

1. Add a length of 0.002740 m to a length of 7.22 mm in millimeter measurement.
2. Subtract a mass of 3 kg from a mass of 4.7 kg.
3. Add three forces of 200 N, 25.333 N, and 38.09 N to have a resultant force.

SOLUTION

1. First, convert 0.002740 m to 2.740 mm.

   Then, identify that 2.740 mm has three significant digits and 7.22 mm has two significant digits to the right of the decimal point, respectively. The least number of significant digits to the right of the decimal point of both lengths in operation is two.
   Next, add 2.740 mm to 7.22 mm, that is,

   $$2.740 \, mm + 7.22 \, mm = 9.960 \, mm$$

   Last, keeping two significant digits to the right of the decimal point in the result, the resulted length is **9.96 mm, which has three significant digits.**

2. First, identify that 3 kg has no significant digit and 4.7 kg has one significant digit to the right of the decimal point. The least number of significant digits to the right of the decimal point of both masses in operation is zero.

   Then, subtract 3 from 4.7 kg, that is,

   $$4.7 \, kg - 3 \, kg = 1.7 \, kg$$

   Last, no significant digit is kept to the right of the decimal point in the result, and the resulting mass is **2 kg, which has 1 significant digit.**

3. First, identify that 200 N has no significant digit, 25.433 N has three significant digits, and 38.09 N has two significant digits to the right of the decimal point, respectively. The least number of significant digits to the right of the decimal point of the forces in operation is zero.

Then, add three forces together, that is,

$$200\,N + 25.333\,N + 38.09\,N = 263.423\,N$$

Last, no significant digit is kept to the right of the decimal point in the result; the resulting force is **263 N, which has three significant digits.**

Exercise 6.12 (multiplication and division)

Find the final numerical value and determine the number of significant digits for the following calculations.

1. Multiplying a length of 2,000 mm with a width of 11.20 m to find the area in meter measurement
2. Multiplying a mass of 0.097 kg with an acceleration of 3.24 m/s² to find the resultant force
3. Dividing a mass of 1,240 kg by a volume of 3.71 m³ to find the specific volume

SOLUTION

1. First, convert 2,000 mm to 2 m.

   Then, identify that 2 m has no significant digit and 11.20 m has two significant digits to the right of the decimal point, respectively. The least number of significant digits to the right of the decimal point of both lengths in operation is zero. Next, multiply 2 m to 11.20 m, that is,

$$2\ m \times 11.20\ m = 22.40\ m^2$$

   Last, no significant digit is kept to the right of the decimal point in the result; the area is **22 m², which has two significant digits.**

2. First, identify that 0.097 kg has two significant digits and 3.24 m/s² has two significant digits to the right of the decimal point, respectively. The least number of significant digits to the right of the decimal point of two numbers in operation is two.

   Then, multiply 0.097 kg by 3.24 m/s², that is,

$$0.97\,kg \times 3.24\frac{m}{s^2} = 0.314,28kg\frac{m}{s^2} = 0.314,28N$$

   Last, keep two significant digits to the right of the decimal point in the result, the force is **0.31 N, which has two significant digits.**

3. First, identify that 1,240 kg has no significant digit and 3.71 m³ has two significant digits to the right of the decimal point. The least number of significant digits to the right of the decimal point of two numbers in operation is zero.

Then, multiply 1,240 kg to 3.71 m³, that is,

$$1,240\,\text{kg} \div 3.71\text{m}^3 = 334.231,8\,\frac{\text{kg}}{\text{m}^3}$$

Last, no significant digit is kept to the right of the decimal point in the result; the specific volume is **334 kg/m3, which has three significant digits.**

In general, three significant digits to the right of the decimal point in engineering practice are sufficiently accurate. However, fewer or more significant digits may be applicable depending on application circumstances.

Exercise 6.13 (FE style)

Determine which outcomes from the calculations below have two significant digits to the right of the decimal point.

(A) 30.75 m² + 2.45 m × 3.12 m
(B) 3.7 N + 4.56 N − 2.161 N
(C) 4.35 lbm × 2.81 lbm
(D) 101.548 kg/81.302 kg + 4.392 kg

SOLUTION

The correct answers are **(A)** and **(C)** since the least number of significant digits to the right of the decimal point of the values in both operations are two.

6.3 ROUNDING OFF NUMBERS

Rounding off a number means keeping a numerical value to the required number of significant digits. The digits in the numerical value after the right side of the number of significant digits can be discarded. The rounded number, therefore, is an approximate value of the actual number in a shorter and more explicit representation but without losing accuracy. In rounding off numbers, the terms *number of significance digits*, *place value*, and *face value* are frequently encountered.

Place Value and Face Value

The place value and the face value of a number are not the same. They may be confused with each other.

- The place value describes the digit value in a given number on the basis of its position.
- The face value represents the digit itself at the position in the given number.

The following table illustrates the meaning of the place value and the face value of the number 2,596.

Digit in 2,596	Place value	Face value
2	2 Thousands ($2 \times 1,000 = 2,000$)	2
5	5 Hundreds ($5 \times 100 = 500$)	5
9	9 Tens ($9 \times 10 = 90$)	9
6	6 units ($6 \times 1 = 6$)	6

Table 6.1 shows an illustration of the number of significant digits, the place value, and the face value of the number 24,136,075.984,1.

Exercise 6.14

Identify the number of significant digits, the place values, and the face values of 3 and 7 in a number 3,743.

SOLUTION

Referring to Table 6.1,

- the number of significant digits of the number of 3,742 is **4**.
- the place values of 3 and 7 in number 3,743 is **3,000 and 700**, respectively.
- the face value of 3 and 7 in 3,743 is **3 and 7**, respectively.

When rounding off, numbers can be rounded up or rounded down. This depends on the face value of the neighboring digit being rounded. There are two sets of rules to keep rounded numbers as accurate as possible. One is for rounding off whole numbers, and another is for rounding off decimals.

Rounding Off Whole Numbers

When rounding a whole number, numbers can be rounded to the nearest ten, the nearest hundred, the nearest thousand, and so on. The rounding-off steps follow:

- Determine the face value to which the number needs to be rounded.
- Look at the number to the right of the number being rounded.
  Rule 1: If that number is equal to 4 or smaller, keep the desired place value as it is.
  Rule 2: If that number is equal to 5 or larger, add 1 to the desired place value.
- Change all the values to the right of the desired place value to zeros.
  Rule 3: When the number to be treated is 5, if the number being rounded is an even number, then add 1 to the rounded number. If the number being rounded is an odd number, then drop 5 without adding to the rounded number.
  Rule 4: If the being rounded digit is zero, zero is treated as an even number to use Rule 3.

TABLE 6.1

The Number of Significance Digits, the Place Value, and the Face Value of a Number

Number 24,136,075.984,1 (All digits are significant in the number)

Place value	Ten Millions (10,000,000)	Millions (1,000,000)	Hundreds Thousands (100,000)	Ten Thousands (10,000)	Thousands (1,000)	Hundreds (100)	Tens (10)	Ones (1)	Decimal Point	Tenths (1/10)	Hundredths (1/100)	Thousandths (1/1,000)	Ten Thousandths (1/10,000)
Face value	2	4	1	3	6	0	7	5	.	9	8	4	1

←— Larger value   Smaller value —→

Exercise 6.15

Rounding a length of 827,642 mm to the nearest tens and thousands place in millimeters, respectively.

SOLUTION

Referring the steps described earlier,

1. the face values in the tens place and thousands place are 4 and 7, respectively.
2. the numbers to the right of the tens place and thousands place are 2 and 6.
3. since 2 is smaller than 5 and 6 is larger than 5, 2 is dropped in rounding to tens place and 1 is added to the thousands place, $1 + 7 = 8$.
4. Changing all the values to the right of the desired place value to zeros in rounding to tens place and thousands place.

Therefore, given length of 827,642 mm, **the rounded digit nearest tens is 827,640 mm and nearest thousands is 828,000 mm.**

Exercise 6.16

Rounding a pressure of 308,563 Pa to four significant digits and five significant digits, respectively.

SOLUTION

Referring to the steps described earlier,

1. the face value of the four significant digits and the five significant digits is 5 and 6.
2. the digit to the right of four significant digits and five significant digits is 6 and 3.
3. since 6 is larger than 5 and 3 is smaller than 5, 1 is added to the digit of four significant digits, $1 + 5 = 6$ and 3 is dropped in rounding to five significant digits.
4. change all the digits to the right of the desired significant digits to zero in rounding to four significant digits and five significant digits.

Therefore, given the pressure of 308,563 Pa, **the rounded pressure to four significant digits is 308,600 Pa and five significant digits is 308,560 Pa.**

Exercise 6.17

Estimate the number of significant digits of a mass of 3,553 kg and round the mass to the first face value 5 and the second face value 5 counting from the left side, respectively.

SOLUTION

The number of significant digits of the mass of 3,553 kg is four. The numbers are 3, 5, 5, and 3, respectively.

Referring to the steps described earlier,

1. the face value to the right of the first face value 5 and the second face value 5 is 3 and 5, respectively.
2. since 3 to the right of the first face value 5 is smaller than the first face value 5, 3 is dropped.
3. since the second face value 5 is an odd digit, 5 to the right of the second face value 5 is dropped.
4. change all the digits to the right of the desired significant digits to zero in rounding to the first face value 5 and the second face value 5, respectively.

Therefore, given the mass of 3,553 kg, **the rounded mass to the first face value 5 is 3,550 kg and to the second face value 5 is 3,500 kg.**

Exercise 6.18 (FE style)

Determine which outcomes resulting from the following expressions have the face value 5 at the hundreds place after rounding the outcomes to the hundreds place value.

(A) 5,041 N + 2,453 N
(B) 7,126 lbm − 2,0.35 lbm
(C) 415 W × 11 W
(D) 5,082 psia/2 psia

SOLUTION

The correct answers are **(A)** and **(D)**. For (A), the outcome is 7,494 N, and the rounded number is 7,500 N. For (B), the outcome is 5,091 lbm, and the rounded number is 5,100 lbm. For (C), the outcome is 4,565 W, and the rounded number is 4,600 W. For (D), the outcome is 2,541 W, and the rounded number is 2,500 W.

Rounding Off Decimals

The procedure of rounding off decimals are almost the same as that of rounding whole numbers. The only difference is that instead of rounding to tens, hundreds, thousands, and so on, rounding off decimals is to tenths, hundredths, thousandths, and so on. The rounding-off steps are the following:

• Determine the face value to which the decimal number will be rounded.
• Look at the number to the right of being rounded.
  Rule 1: If that number is equal to 4 or smaller, keep the desired place value as it is.
  Rule 2: If it is equal to 5 or larger, add 1 to the desired place value.
• Remove all the values to the right of the desired place value.
  Rule 3: When the number to be treated is 5, if the being rounded digit is an even number, then add 1 to the rounded number. If the being rounded digit is an odd number, then drop the 5 without adding to the rounded number.

Rule 4: If the being rounded number is zero, zero is treated as an even number to use Rule 3.

Exercise 6.19

Determine the number of significant digits of a gas density of 0.054,83 kg/m³. Round off the gas density to two and three significant digits, respectively.

SOLUTION

The number of significant digits of the gas density of 0.054,83 kg/m³ is four. The numbers are 5, 4, 8, and 3. Referring to the steps described earlier,

1. the digits of the two significant digits and three significant digits are 4 and 8.
2. the digits to the right of the two significant digits and three significant digits are 8 and 3.
3. since 3 is smaller than 5 and 8 is larger than 5, 3 is dropped in rounding to the three significant digits and 1 is added to 4 in rounding to the two significant digits, $1 + 4 = 5$.
4. remove all the values to the right in rounding to two and three significant digits.

Therefore, given the gas density of 0.054,83 kg/m³, **the rounded density to two significant digits is 0.055 kg/m³ and three significant digits is 0.0548 kg/m³.**

Exercise 6.20

Determine the number of significant digits of the weight 26.07 N and round off the weight to the 0.1 place.

SOLUTION

The number of significant digits of the weight of 26.07 N is four. The numbers are 2, 6, 0, and 7. Referring to the steps described earlier,

1. the digit in the 0.1 place is 0.
2. the digit to the right of the 0.1 place is 7.
3. since 7 is bigger than 5 and the being rounded digit in the 0.1 place is zero, 1 is added to 0 in rounding to the 0.1 place, $1 + 0 = 1$.
4. Remove all the values to the right in rounding to the 0.1 place.

Therefore, given the weight of 26.05 N, **the rounded weight to the 0.1 place is 26.1 N.**

A summary of rounding off the numerical value for a quantity of 706.636 kg is illustrated in Table 6.2.

TABLE 6.2
Summary of Rounding Off a Numerical Value for a Quantity of 706.636 kg

Number of Significant Digits	Rounded Value	Rule
6	706.636	All digits are significant
5	706.64	6 rounds the 3 to 4 since 6 > 5
4	706.6	4 dropped since 3 < 5
3	707	6 rounds the 6 to 7 since 6 > 5
2	710	7 rounds the 0 to 1 since 7 > 6
1	700	1 dropped since 1 < 5

Exercise 6.21 (FE style)

Determine the outcomes of the following expressions that have <u>more than four</u> significant numbers for the numerical values after rounding the outcomes to thousandths place value.

(A) $12.74\,\text{mm} \times 0.826\,\text{mm}$

(B) $19.151\dfrac{\text{kg}}{\text{m}^3} + 1.2084\dfrac{\text{kg}}{\text{m}^3}$

(C) $4.34\,\text{lbf} - 2.1381\,\text{lbf}$

(D) $2.022626\,\text{N} / 1.01\,\text{m}^2$

SOLUTION

The correct answers are **(A)** and **(B)**. For (A), the outcome is 10.52423 mm², and the rounded number is 10.524 mm². For (B), the outcome is 20.3594 kg/m³, and the rounded number is 20.259 kg/m³. But for (C), the outcome is 2.202 lbf, and the rounded number is 2.202 lbf, which has only four significant numbers, and for (D), the outcome is 2.0026 Pa, and the rounded number is 2.003 Pa, which has only four significant numbers.

Rounding Order of Sums

Rounding can make sums easy. But the order matters for the result accuracy when calculating is first and then rounding or rounding is first and then calculating. For example, determine a power sum of 4.6 kW and 3,800 W in kilowatts to the ones place. Converting W to kW by using the unit conversion factor,

$$1,000\,\text{W} = 0.001\,\text{kW}$$

therefore,

$$3,800\,\text{W} = 3,800 \times 0.001\,\text{kW} = 3.8\,\text{kW}$$

- Calculate first and then round:

$$4.6 \text{ kW} + 3.8 \text{ kW} = 8.4 \text{ kW} = 8 \text{ kW}$$

- Round first and then calculate:

$$4.6 \text{ kW} + 3.8 \text{ kW} = 5 \text{ kW} + 4 \text{ kW} = 9 \text{ kW}$$

Both results are acceptable by the calculations in terms of rounding off. However, comparing their accuracies in percentage of difference,

$$\Delta\% = \frac{8.4 - 8 \text{ kW}}{8.4 \text{ kW}} = 0.04762 = 4.762\%$$

$$\Delta\% = \frac{9 - 8.4 \text{ kW}}{8.4 \text{ kW}} = 0.0714 = 7.143\%$$

It can be seen that the way of calculating first and then rounding seems more accurate. Which order should be used depends on the engineering application. Sometimes if the calculation needs a quick estimation and the result accuracy is not so critical, it may be more convenient to round off first and then calculate.

Exercise 6.22

Four electric parts are purchased in an appliance store. The costs are $3.25, $5.88, $6.99, and $12.59, respectively. Estimate the total cost for the parts to have a quick result and find the accuracy in percentage.

SOLUTION

The actual total cost of four electric parts is

$$\$3.25 + \$5.88 + \$6.99 + \$12.59 = \$28.71$$

Since the result is not critical, rounding first and then adding is more convenient:

$$\$3.25 = \$3$$

$$\$5.88 = \$6$$

$$\$6.99 = \$7$$

$$\$12.59 = \$13$$

The estimated total cost is determined to be

$$\$3 + \$6 + \$7 + \$13 = \mathbf{\$29}$$

The accuracy in percentage is

$$1 - \frac{\$29 - \$28.71}{\$29} = 1 - 0.01 = 0.99 = \mathbf{99\%}$$

Exercise 6.23 (FE style)

Mr. Lee and Mr. Cox are calculating the pressure on a plate which requires the outcome in an integer. Their measurement finds that the force on the plate is 405.4 kN and the plate area is 20.1 m². Mr. Lee proposes rounding first and then calculating, while Mr. Cox suggests calculating first and then rounding. Decide which outcome is correct.

(A) Mr. Lee's
(B) Mr. Cox's
(C) Can't decide
(D) No difference

SOLUTION

The correct answer is **(D)** since by rounding first and then calculation, the outcome is

$$\text{Force on the plate: } 405.4 \text{ kN} = 405 \text{ kN}$$

$$\text{the plate area: } 20.1 \text{ m}^2 = 20 \text{ kN}$$

The pressure is

$$405 \text{ kN}/20 \text{ m}^2 \doteq 20.25 \text{ kPa}$$

Rounding the pressure to an integer, the pressure is 20 kPa. While by calculating first and then rounding, the outcome is

$$405.4 \text{ kN}/20.1 \text{ m}^2 = 20.17 \text{ kN/m}^2$$

Rounding the pressure to an integer, the pressure is also 20 kPa. **The outcomes have no difference.**

Notice the different phrasings in rounding off. One says to round off a number to a specific digit position, the number of significant digits, the place value, and the face value in rounding are equivalent if the position is specified. The resulting numbers after rounding are the same.

Exercise 6.24

Determine a mass of 8.534 kg rounded to a 0.01 place by applying (a) the number of significant digits, (b) the place value, and (c) the face value.

SOLUTION

(a) In terms of the number of significant digits, two significant digits after the decimal point are kept corresponding to the 0.01 place in the rounded mass. The digit of two significant digits after the decimal point in 8.534 is 3. Since the digit of the right to two significant digits is 4, which is smaller than 5, the 4 is dropped, and the rounded mass is **8.53 kg**.

(b) In terms of the place value, the hundredths place value after the decimal point are kept corresponding to the 0.01 place in the rounded mass. The digit of the hundredths place value in 8.534 is 3. Since the digit of the right to the second place is 4, which is smaller than 5, the 4 is dropped, and the rounded mass is **8.53 kg**.

(c) In terms of the face value, the face value after the decimal point is 3 corresponding to the 0.01 place in the round mass. Since the digit of the right to the face value is 4, which is smaller than 5, 4 is dropped, and the rounded mass is **8.53 kg**.

Therefore, rounding the mass of 8.534 kg to the specified 0.01 place, the numerical values of the rounded mass are identical from the meaning of the number of significant digits, the place value, and the face value.

7 Dimensioning Objects on Engineering Drawings

An engineering drawing is a type of technical drawing used to provide construction information about objects (or systems, see Chapter 8). Engineering drawings of objects communicate not only geometry shapes but also dimensions of the objects. A dimension is a numerical value expressed in appropriate units of measurement and used to define the size, location, orientation, and geometric characteristics of an object (or a system). The process of adding dimensions to an object on a drawing is known as dimensioning the object. The dimensions on drawings are the key information for construction by using standardized and mathematical verified rules. Engineers in design should ensure the production team can get all information and data directly from the drawing. Accurate and clear dimensioning on drawings helps make manufacturing correctly and efficiently. Inaccurate or improper dimensioning may lead to poor quality, time wasting, and monetary loss. A complete set of dimensions on a drawing permits only one interpretation needed for successful manufacturing. Dimensioning should follow these guidelines:

- Accuracy: Correct values must be given.
- Clearness: Dimensions should be placed in appropriate positions.
- Completeness: Nothing is left out and duplication is not acceptable.
- Readability: Standard dimensioning format must be applied.

7.1 BASICS OF DIMENSIONING

In engineering practice, an object can be presented by an isometric drawing in a three-dimensional view or an orthographic drawing in full two-dimensional views: front, top, bottom, right, and left, as shown in Figure 7.1.

Dimensions in engineering drawings are numerical values with lines, leaders, symbols, and notes, among others, as shown in Figure 7.2(a). Dimensions are generally detailed on the orthogonal drawing in a two-dimensional view, commonly using the front and top views, as shown in Figure 7.2(b). The dimension for a feature is usually only specified in one place once on the drawing to avoid redundancy and the possibility of inconsistency. Dimensioning objects on the isometric drawing in three-dimensional views is acceptable, as shown in Figure 7.2(c). But it is not encouraged unless the object information cannot be drawn from any view using the orthogonal drawing.

DOI: 10.1201/9781003508977-7

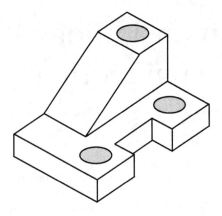

(a) Isometric drawing in three-dimensional view

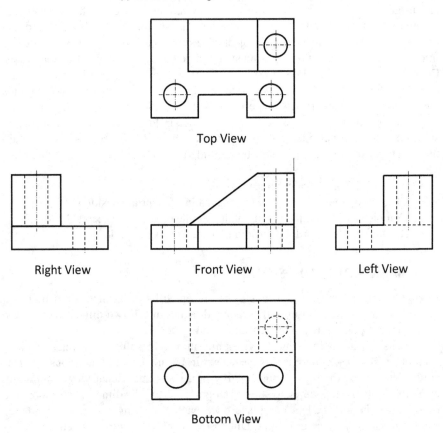

Top View

Right View   Front View   Left View

Bottom View

(b) orthographic drawing in two-dimensional views

FIGURE 7.1   Isometric and orthogonal drawings.

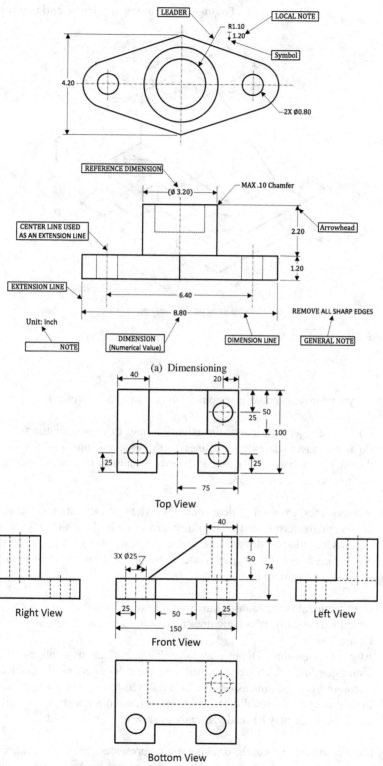

(a) Dimensioning

(b) Dimensioning on two-dimensional views

FIGURE 7.2 Dimensioning.

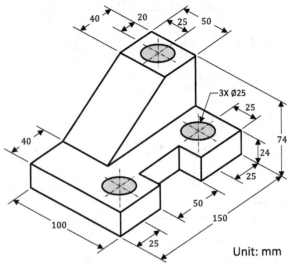

(c) Dimensioning on three-dimensional view

FIGURE 7.2   (Continued)

Some typical mistakes in dimensioning objects should be avoided:

- Do not use object outlines as the extension lines to show dimensions.
- Do not connect the extension lines to the object/feature outlines. A gap should be kept between the end of the extension lines and the object/feature outlines, as shown in Figure 7.3.

Dimension lines are used to determine the extent and direction of dimensions. They are normally terminated by uniform arrowheads. In a drawing of a two-dimensional view, placing dimensions on the views where they can be most easily understood is preferred. There are two methods approved by the ANSI guidance to place dimensions on a drawing:

- Unidirectional dimension—Dimensions have all numbers and notes presented horizontally, which are presented in the same direction, as shown in Figure 7.4(a).
- Aligned dimension—Dimensions have all numbers and notes aligned with a dimension line, which are presented from the sides or edges of an object or a feature. Aligned dimensions are also referred to as pictorial dimensions. In a drawing in aligned dimensions, some dimensions may be read horizontally, and some may be read vertically as shown in Figure 7.4(b).

Only one method throughout the drawing can be presented, either the unidirectional dimension or the aligned dimension. The unidirectional method is commonly preferred because it has better readability. This book uses the unidirectional method.

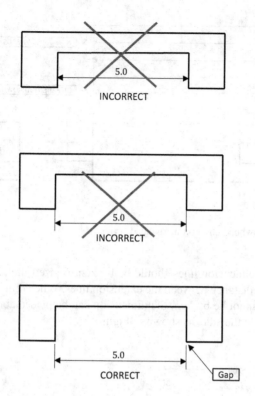

FIGURE 7.3 Two typical mistakes in dimensioning.

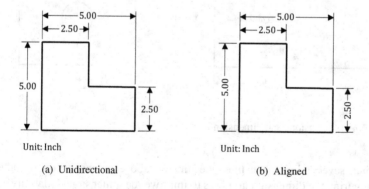

FIGURE 7.4   Unidirectional and aligned dimensioning.

Some other basics of dimensioning drawings include the following:

- Dimensioning to hidden lines should not happen unless necessary.
- A single style of arrowhead should be used throughout the drawing. However, different arrowhead styles may be used; even non-arrowhead styles are applicable when drawing space is limited, as shown in Figure 7.5.

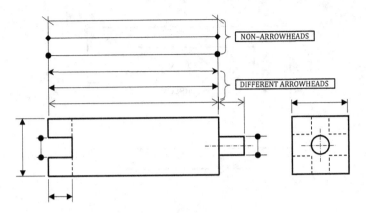

FIGURE 7.5   Arrowhead and non-arrowhead styles.

- Preferably, dimension lines should be broken to insert the dimensions that indicate the distance between the extension lines. When dimension lines are short and cannot be broken, the dimension can be placed above the dimension line or to the side, as shown in Figure 7.6.

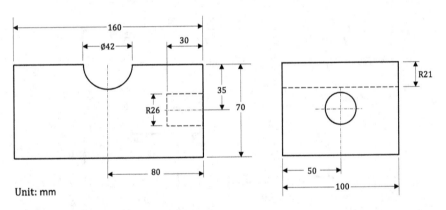

Unit: mm

FIGURE 7.6   Dimensions and dimension lines.

- When several dimension lines are directly above or next to one another, staggering the dimension numbers to improve the dimension readability is a good practice, as shown in Figure 7.7.
- When the space between the extension lines is too small to permit placing the dimension line completely with arrowheads and dimension, the alternative methods of placing dimension lines, the extension lines, and dimension can be used as shown in Figure 7.8.
- Every effort should be made to avoid crossing dimension lines, as shown in Figure 7.9(a), and place the shortest dimension closest to the outlines of the object and the feature, as shown in Figures 7.9(b) and 7.9(c).

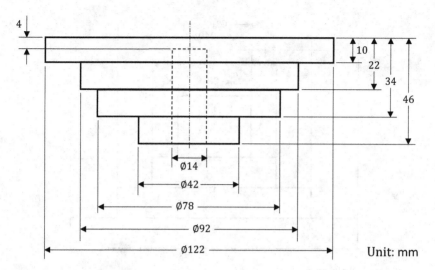

FIGURE 7.7 Stagger dimension number.

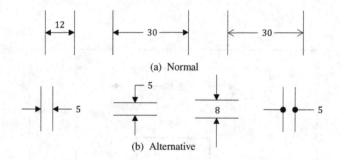

FIGURE 7.8 Alternative methods of dimensioning.

- When the termination for a dimension is not included, for example, when a drawing shows a partial or sectional view, the dimension lines should be extended beyond the center line of the object and feature being dimensioned and presented with only one arrowhead as shown in Figure 7.10.
- Dimension lines should be placed outside the view whenever possible and should be extended to extension lines rather than visible lines. However, when readability can be improved by avoiding either extra-long extension lines or the crowding of dimensions, placing dimensions on views is acceptable, as shown in Figure 7.11.

7.1.1 UNITS OF MEASUREMENT

Although the metric units (the SI) of dimensioning have become the official international standard or measurement, the United States still uses a dual system. Drawings

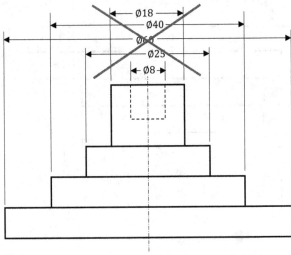

Unit: mm

(a) Incorrect

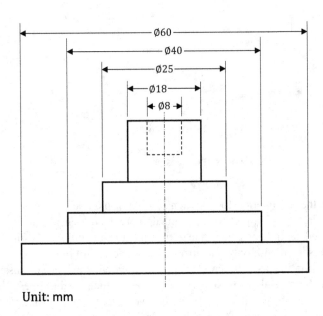

Unit: mm

(b) Horizontal

FIGURE 7.9 Shorter dimensions closing to object and feature outlines.

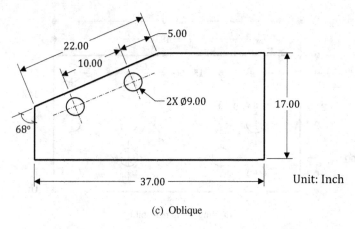

(c) Oblique

FIGURE 7.9 (Continued)

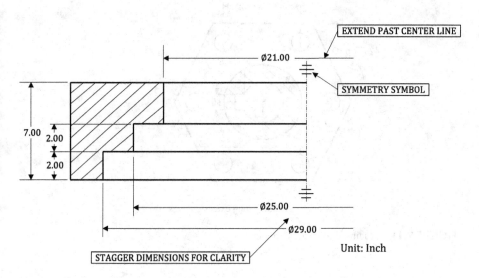

FIGURE 7.10 Termination and partial view dimensioning.

in the United States are dimensioned by either inches/feet in the English system or millimeters in the SI. Keep the following in mind:

- All dimensions on one drawing should be used either in inches/feet or in millimeters/meters. Mixing unit systems or measurements should be avoided.
- The drawing should contain a note indicating the unit or measurement used on the drawing. The note is a general note.
- The inch marks (″) does not need to be shown in dimensions on drawings.

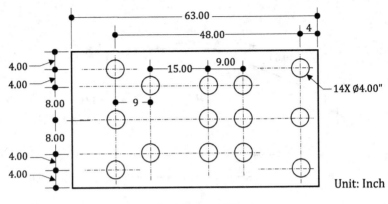

(a) Improving readability

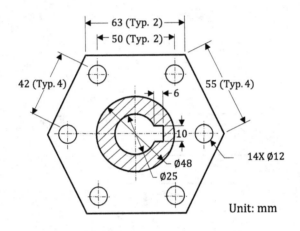

(b) Avoiding long extension lines

FIGURE 7.11   Placing dimensions on views.

7.1.1.1 Inch Units
Decimal-Inch Style

The decimal-inch style is for objects to be designed with dimensioning in basic decimal increments, preferably 0.02 in., and is expressed with a minimum of two digits to the right of the decimal point in linear dimensions, as shown in Figure 7.12. If a 0.02 in. style is used, the second decimal place (hundredths) is always either an even number or zero.

- If design modules in increments of 0.02 are applied, such as .04, .06, and .16, the dimensions can be halved for center distances without it being necessary to increase the number of decimal places.

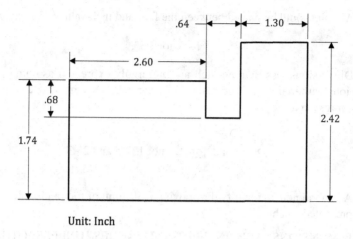

Unit: Inch

FIGURE 7.12 Dimensioning of decimal-inch style.

- If the design modules of incremental 0.02 in are not applied, such as .01, .03, and .11, the dimensions should be used only when needed to meet design requirements. When greater accuracy is required, dimensions may be able to be expressed as a three- or four-place decimal number, for example, 0.725 and 1.875.
- Whole dimensions are complete when shown with a minimum of two zeros to the right of the decimal point for integers. For example, inches of 12 and 28 should be

$$12.00, \qquad not \qquad 12$$

and

$$28.00, \qquad not \qquad 28.$$

- An inch value of less than 1 is not necessary to show a zero to the left of the decimal point. For example, inches of 0.23 and 0.52 should be

$$.23 \qquad not \qquad 0.23$$

and

$$.52 \qquad not \qquad 0.52$$

Foot-and-Inch Style

The dimensions of foot-and-inch style are broadly applied in system drawings and drawings of large objects:

- All dimensions equal to or larger than 12 in. are specified in feet and inches. For example, dimensions of 36 in. and 52 in. are expressed as

$$3'\text{-}0 \text{ and } 4'\text{-}4, not \; 3' \text{ and } 4'\text{-}4'',$$

respectively.

- A dash should be placed between the foot and inch values. For example,

$$1'-5 \text{ not } 1'5''.$$

- Dimensions less than an inch are preferably expressed as common fractions rather than as decimals. For example, 48.5 in. and 24.75 in. are expressed as

$$12'-\frac{1}{2} \text{ and } 12'-\frac{3}{4} \text{ not } 12'-.5 \text{ and } 2'-.75$$

- A general note to identify the drawing in the unit of foot-and-inch should be presented, such as

DIMENSIONS ARE IN FEET-AND-INCHES UNLESS OTHEWISE SPEFICIED

or

UNIT: UNIT OF DIMENSION.

Figure 7.13 shows a typical dimensioning of the foot-and-inch style.

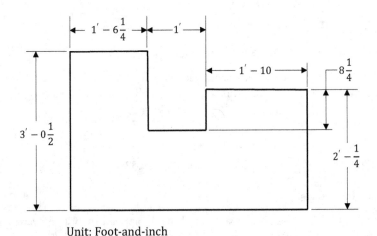

Unit: Foot-and-inch

FIGURE 7.13 Dimensioning of the foot-and-inch style.

7.1.1.2 Metric Units

The metric units in the SI on engineering drawings are the millimeter (mm) for linear measure:

- Whole numbers from 1 to 9 should be shown without a zero to the left of the number or a zero to the right of the decimal point. For example, 2 mm and 4 mm should be expressed as

$$2, \qquad \text{not 02 or 2.0,}$$

and

$$4, \qquad \text{not 04 or 4.0.}$$

- A millimeter value of less than 1 should be shown with a zero to the left of the decimal point, which is the purpose to distinguish it from the inch unit in the English system. For example, 0.3 mm and 0.46 should be specified as

$$0.3, \qquad \text{not .3 or .30,}$$

and

$$0.46, \qquad \text{not .46.}$$

- A general note to identify a metric drawing should be presented, such as

DIMENSIONS ARE IN MILLIMETERS UNLESS OTHEWISE SPECIFIED

or

UNIT: UNIT OF DIMENSION.

Figure 7.14 shows a typical dimensioning of the metric units.

7.1.1.3 Dual Dimensioning

Dual dimensioning is a dimension style that shows both inches and millimeters for one dimension on a drawing. The style has advantages for engineers to communicate using

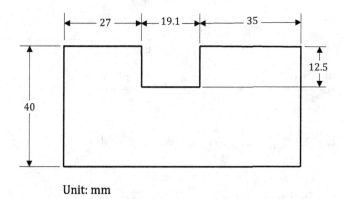

Unit: mm

FIGURE 7.14 Dimensioning of the metric unit.

different unit systems. Many internationally operated companies preferably adopt the dual system of dimensioning. The formats of the dual dimensioning can be in

- Position method: $\dfrac{\text{MILLIMETER}}{\text{INCH}}$ or MILLIMETER/INCH

- Bracket method:

$$\begin{array}{c}\text{[INCH]}\\ \text{MILLIMETER}\end{array}\text{ or MILLIMETER [INCH]}$$

Figure 7.15 shows the dual dimensioning on drawings.

7.1.1.4 Units Common to Either System

Some particular measurements can satisfy both units of the English system and the SI. For example, tapers such as .4 in per inch and 0.4 mm per millimeter can be expressed by a ratio with a taper symbol,

$$0.4{:}1$$

or a general note,

$$\text{TAPER } 0.4{:}1.$$

Angular dimensions can also be specified the same way in inches and millimeters on drawings as shown in Figure 7.16.

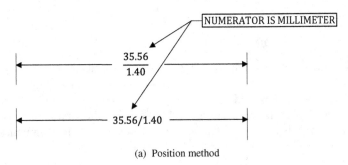

(a) Position method

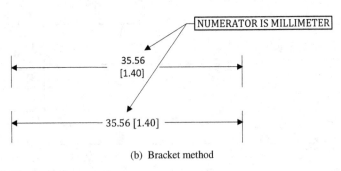

(b) Bracket method

FIGURE 7.15   Dual dimensioning.

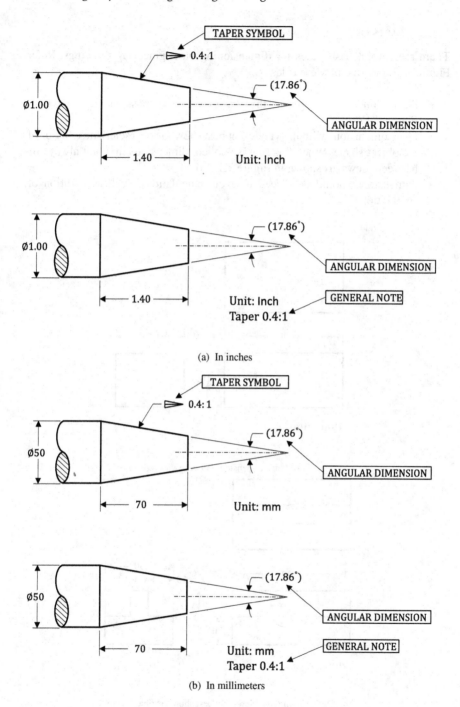

FIGURE 7.16 Units common to either the English system or the SI.

7.1.2 RULES OF DIMENSIONING

There are several basic rules for dimensioning on engineering drawings. Refer to Figure 7.17 for some of these rules.

- Place dimensions between the views whenever possible, as shown in Figure 7.17(a).
- Place dimensions with the views that best show the characteristic contour of the object shape. When this rule is applied, dimensions may not always be between views, as shown in Figure 7.17(b).
- Dimensions should be placed to avoid calculations, such as addition or subtraction.

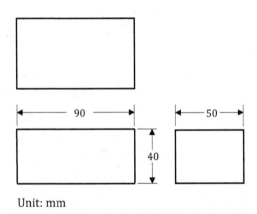

(a) Place dimensions between views

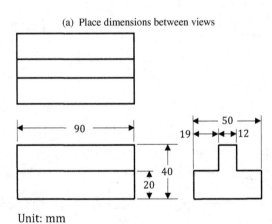

(b) Dimensioning the views that best show the object shape

FIGURE 7.17   Basic rules of dimensioning on engineering drawings.

7.1.3 SYMBOLS AND ABBREVIATIONS

The symbols and abbreviations are frequently used in engineering drawings to save space and time, particularly when a large or complicated drawing is crowded. The symbols and abbreviations applied should be as simple as possible and have unique clear meanings. Table 7.1 lists some commonly used symbols and abbreviations on engineering drawings.

TABLE 7.1
Commonly Used Symbols and Abbreviations

Symbols			
Meaning		**Meaning**	
And	&	Dimension Origin	⦶
Arc Length	⌢33	None	None
Copper	Cu	Number	NO.
Center Line	℄	Perpendicular	⊥
Conical Slope (Taper)	⊳	Radius	R
Counterbore	⌴	Reference Dimension	( )
Countersink	∨	Repeated Dimension	4X
Degree (Angle)	°	Spherical Diameter	SØ
Depth	↧	Spherical Radius	SR
Diameter	Ø	Symmetrical	≡

Abbreviations			
Meaning		**Meaning**	
Approximate	APPROX	Liter	L
Assembly	ASSY	Machine Steel	MST
Brass	BR	Malleable Iron	MI
Carbon Steel	CS	Material	MATL
Cast Iron	CI	Maximum	MAX
Center Line	CL	Meter	m
Center to Center	C to C	Millimeter	mm
Centimeter	cm	Minimum	MIN
Concentric	CONC	Module	MDL
Counterdrill	CDRILL	Newton	N
Countersink	CSK	Nominal	NOM
Datum	DAT	Not To Scale	NTS
Degree (Angle)	DEG	Outside Diameter	OD
Depth	DP	Parallel	PAR
Diameter	DIA	Pascal	Pa
Dimension	DIM	Perpendicular	PERP
Drawing	DWG	Plate	PL

(*Continued*)

TABLE 7.1 (*Continued*)
Commonly Used Symbols and Abbreviations

Symbols

Meaning		Meaning	
Eccentric	ECC	Radius	R
Equally Spaced	EQL SP	Reference or Reference Dimension	REF
Figure	FIG	Revolutions per Minute	RPM
Foot	FT	Right Hand	RH
Gage	GA	Second (Time)	SEC
Head	HD	Socket	SOCK
Heavy	HVY	Spherical	SPHER
Hexagon	HEX	Spherical Radius	SR
Hydraulic	HYDR	Square	SQ
Inch	IN.	Steel	STL
Inside Diameter	ID	Symmetrical	SYM
International Organization for Standardization	ISO	Straight	STR
International Pipe Standard	IPS	Thread	THD
Kilogram	kg	Through	THRU
Kilometer	km	Watt	W
Left Hand	LH	Wrought Iron	WI
Length	LG	Wrought Steel	WS

7.1.4 REFERENCE AND SYMMETRICAL DIMENSIONS

A reference dimension is used to provide information or for inspection only, which is not required to be met for production. The value of the reference dimension is presented in parentheses. Figure 7.18 shows the reference dimension applied to a drawing.

An object is said to be symmetrical when the features on each side of the center line or median plane are identical in size, shape, and location. Partial views are often drawn for the sake of economy or saving space. When only one-half of the configuration of a symmetrically shaped part is presented, symmetry is indicated by applying a pair of the symmetry symbols on the two sides of the object on the center line, as shown in Figure 7.19.

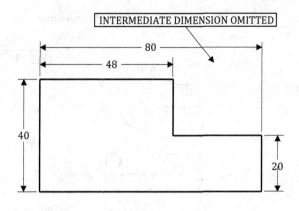

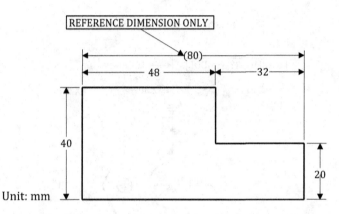

Unit: mm

FIGURE 7.18 Reference dimension.

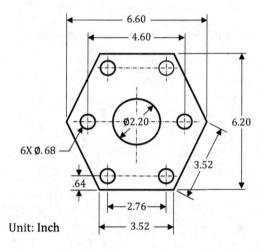

Unit: Inch

(a) Complete drawing

FIGURE 7.19 Symmetrical dimensions.

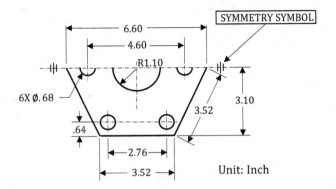

(b) Symmetrical drawing (vertical)

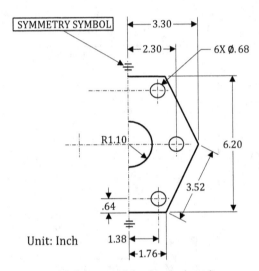

(c) Symmetrical drawing (horizontal)

FIGURE 7.19   (Continued)

Exercise 7.1

The drawing as shown in Figure *Exercise 7.1* contains typical errors of dimensioning. (a) Point out all the dimensioning errors and (b) redraw the drawing with the correct ones.

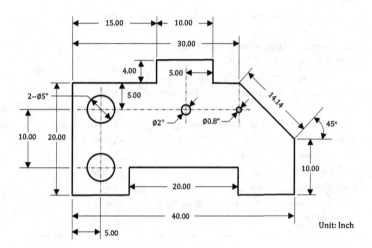

FIGURE *Exercise 7.1.*

SOLUTION

(a) All the dimensioning errors are shown in the following drawing. The corrections are shown in the following table.

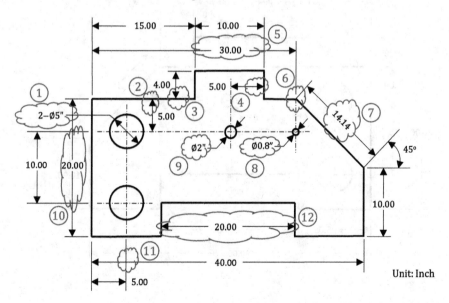

No.	Error	Correction
1	- Improper dimension format - Crossing dimension lines	- Use 2X Φ5.00. - Redraw.
2	- Improperly using the object outline as the extension line	- Redraw.
3	- Improperly using the object outline as the extension line	- Redraw.
4	- Improperly using the object outline as the extension line	- Redraw.
5	- Crossing dimension lines	- Redraw.
6	- Connecting the extension line to the object outline	- Keep a gap between them.
7	- Improperly aligned dimensioning - Dimension should be a reference dimension.	- Use the unidirectional dimension. - Use the reference dimension (14.14).
8	- Improper dimension format	- Use Φ.80.
9	- Improper dimension format	- Use Φ2.00.
10	- Crossing dimension lines	- Redraw.
11	- Crossing dimension lines	- Redraw.
12	- Improperly using the object outlines as the extension lines	- Redraw.

(b) The redrawn drawing is shown with the correct dimensioning.

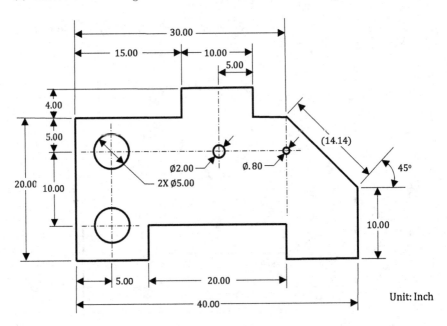

7.2 DIMENSIONING METHODS

In general, the choice of the most suitable dimension style and dimensioning method mainly depends on how the object can be produced and whether the drawing is intended for unit or mass production.

- *Unit production* refers to when each object is to be made separately, which uses general purpose tools and machines.
- *Mass production* refers to when the object is produced in quantity, where special tools and gages are usually provided.

The following dimensioning methods are commonly used for engineering drawings.

7.2.1 RECTANGULAR COORDINATE DIMENSIONING

This is a method to present distance, location, and size by means of linear dimensions measured parallel or perpendicular to reference axes or datum planes that are perpendicular to one another. Dimensioning coordinates with dimension lines must clearly identify the datum features from which the dimensions originate, as shown in Figure 7.20.

Rectangular Coordinates for Arbitrary Points

Using coordinates for arbitrary points of reference without a grid appearing adjacent to each point or in tabular form is shown in Figure 7.21. If the AutoCAD software is used, coordinates at any point will be automatically displayed when the point is selected.

Rectangular Coordinate Dimensioning without Dimension Lines

Dimensions may be shown on extension lines without using dimension lines or arrowheads. The base lines may be zero coordinates in X and Y directions. The feature dimensions can be shown in the drawing or listed in a table, as in Figure 7.22.

Tabular Dimensioning

The tabular dimensioning is a method of coordinate dimensioning in which dimensions from mutually perpendicular planes are listed in a table on the drawing rather than on the pictorial delineation. This method has advantages for drawings that require the locations of a large number of similarly shaped features when objects from numerical control are dimensioned, as shown in Figure 7.23.

For circular objects, having the object dimensioned symmetrically about its center may be convenient, as shown in Figure 7.24. When the center lines are designated as the base (zero) lines, positive and negative values will occur. These values are shown with the dimensions locating the holes.

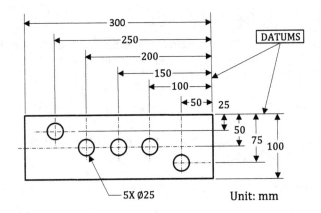

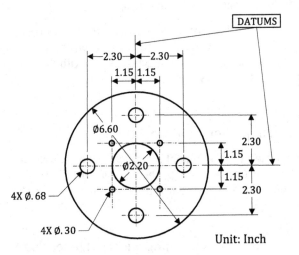

FIGURE 7.20 Rectangular coordinate dimensioning.

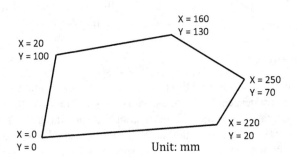

(a) Coordinates for arbitrary points

FIGURE 7.21 Rectangular coordinates for arbitrary points and in tabular form.

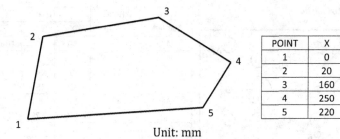

POINT	X	Y
1	0	0
2	20	100
3	160	130
4	250	70
5	220	20

Unit: mm

(b) Coordinates in tabular form

FIGURE 7.21 (Continued)

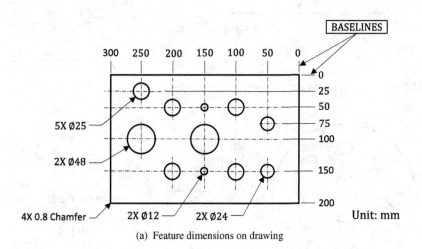

(a) Feature dimensions on drawing

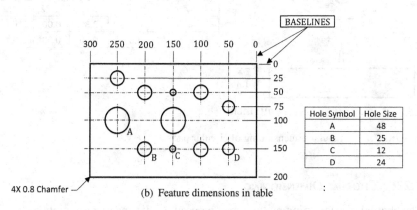

Hole Symbol	Hole Size
A	48
B	25
C	12
D	24

(b) Feature dimensions in table

FIGURE 7.22 Dimensioning without dimension lines.

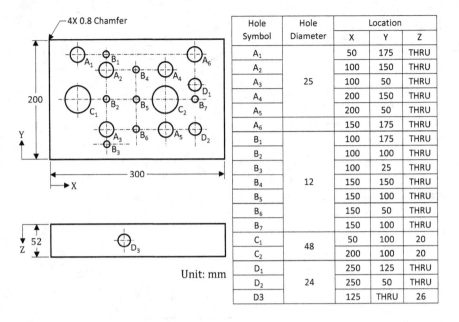

Hole Symbol	Hole Diameter	Location		
		X	Y	Z
A_1		50	175	THRU
A_2		100	150	THRU
A_3	25	100	50	THRU
A_4		200	150	THRU
A_5		200	50	THRU
A_6		150	175	THRU
B_1		100	175	THRU
B_2		100	100	THRU
B_3		100	25	THRU
B_4	12	150	150	THRU
B_5		150	100	THRU
B_6		150	50	THRU
B_7		150	100	THRU
C_1	48	50	100	20
C_2		200	100	20
D_1		250	125	THRU
D_2	24	250	50	THRU
D3		125	THRU	26

FIGURE 7.23 Tabular dimensioning.

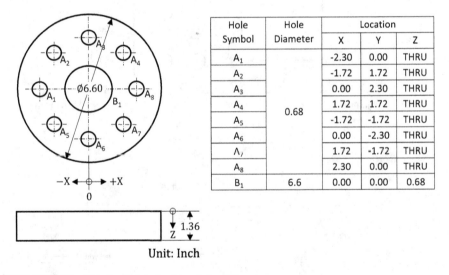

Hole Symbol	Hole Diameter	Location		
		X	Y	Z
A_1		-2.30	0.00	THRU
A_2		-1.72	1.72	THRU
A_3		0.00	2.30	THRU
A_4	0.68	1.72	1.72	THRU
A_5		-1.72	-1.72	THRU
A_6		0.00	-2.30	THRU
A_7		1.72	-1.72	THRU
A_8		2.30	0.00	THRU
B_1	6.6	0.00	0.00	0.68

FIGURE 7.24 Tabular dimensioning of circular parts.

7.2.2 CHORDAL DIMENSIONING

The chordal dimensioning method may be used for the spacing of points on the circumference of a circle relative to a line when manufacturing practices indicate that this will be more convenient, as shown in Figure 7.25.

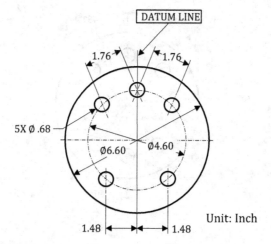

FIGURE 7.25 Chordal dimensioning.

7.2.3 POLAR COORDINATE DIMENSIONING

Polar coordinate dimensioning is commonly used in circular planes or circular configurations of features. This method indicates the position of a point, line, or surface by means of a linear dimension and an angle, other than 90°, which is implied by the vertical and horizontal center lines, as shown in Figure 7.26.

7.2.4 COMMON-POINT DIMENSIONING

When several dimensions originate from a common reference point or line, a method called the common-point or datum dimensioning can be used. Using this method, dimensioning from reference lines may be executed as parallel or a superimposed running dimensioning, as shown in Figure 7.27.

- Superimposed running dimensioning is simplified parallel dimensioning and may be used when there are space problems. Dimensions should be place near the arrowhead, in line with the corresponding extension line, as shown in Figure 7.27(b). The origin is indicated by a circle, and the opposite end of each dimension is terminated with an arrowhead.
- The method may be advantageous to use superimposed running dimensions in two directions. Under this circumstance, the origins may at the center of a hole or other feature as shown in Figure 7.28

7.2.5 CHAIN DIMENSIONING

When a series of dimensions is applied on a point-to-point basis, it is called chain dimensioning, as shown in Figure 7.29. A possible disadvantage of this method is that it may result in an undesired accumulation of tolerances between individual features.

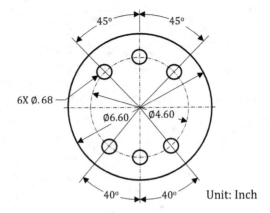

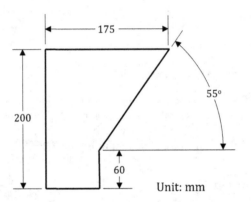

FIGURE 7.26   Polar coordinate dimensioning.

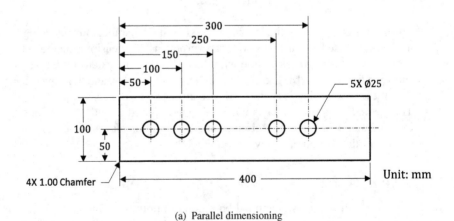

(a) Parallel dimensioning

FIGURE 7.27   Common-point dimensioning.

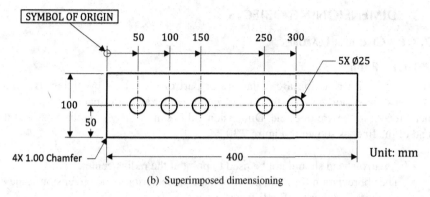

(b) Superimposed dimensioning

FIGURE 7.27 (Continued)

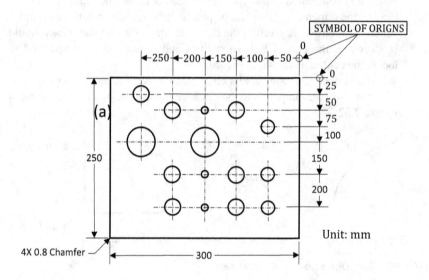

FIGURE 7.28 Superimposed running dimensioning in two dimensions.

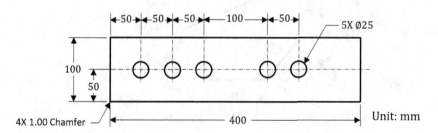

FIGURE 7.29 Chain dimensioning.

7.3 DIMENSIONING OBJECTS

7.3.1 CIRCULAR FEATURES

7.3.1.1 Radii

In general, dimensioning a circular arc or part of a circle is by giving its radius. A radius dimension line points from the radius center and terminates with an arrowhead touching the arc or circle. Dimension value can be shown on the line or at the end of the line, as shown in Figure 7.30.

- An arrowhead should not be used to point at the radius center.
- The abbreviation R (not italic) should be placed prior to the dimension value for units in either the English system or the SI.
- If space is limited, such as a small radius, the radial dimension line can be extended through the radius center.
- When it is convenient to place the arrowhead between the radius center and the arc, the dimension value can be placed either inside or outside the arc.
- When a dimension is given to the center of the radius, a small cross should be drawn at the center. Extension lines and dimension lines are used to locate the center, as shown in Figure 7.31.
- When a tangent line contacts its endpoints on two circles' circumferences, the lines from the endpoints to the circle cents must be parallel, as shown in Figure 7.32.

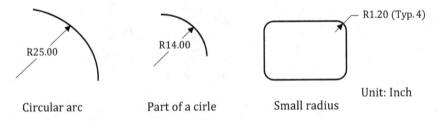

Circular arc   Part of a cirle   Small radius

FIGURE 7.30 Dimensioning of arrowheads.

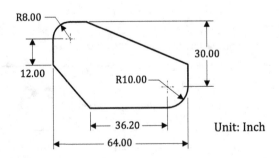

FIGURE 7.31 Dimensioning location.

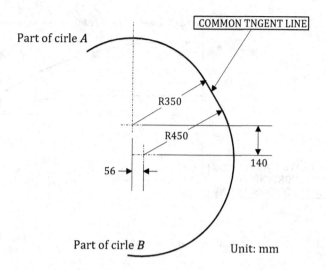

FIGURE 7.32   Radii with a common tangent line.

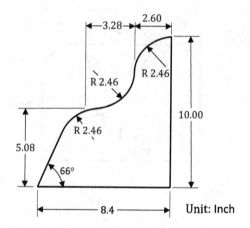

FIGURE 7.33   Radial arcs with tangent lines.

- When the location of the center is not critical, the radial arcs may be located by tangent lines as shown in Figure 7.33.
- When the center of a radius is outside the drawing or interferes with another view, the radius dimension line may be foreshortened. The portion of the dimension line next to the arrowhead should be redial relative to the curved line, as shown in Figure 7.34.
- When the radius dimension line is foreshortened and the center is located by coordinate dimensions, the dimensions locating the center should be shown as foreshortened and a general note of not to scale (NTS) is used, as shown in Figure 7.35.

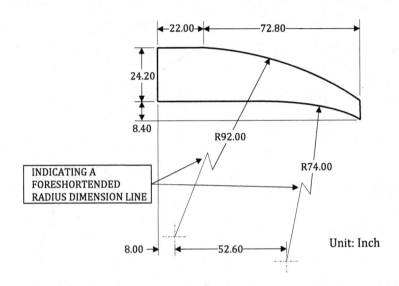

FIGURE 7.34   Foreshortened radius dimension.

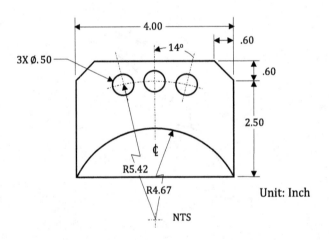

FIGURE 7.35   Foreshortened radius dimension with a note of not to scale (NTS).

- When a radius is dimensioned in a view that does not show the true shape of the radius, TRUE R may be added before the radius dimension, as shown in Figure 7.36.
- Simple fillet and corner radii may also be dimensioned by a general note, such as

ALL ROUNDS AND FILLETS UNLESS OTHERWISE SPECIFIED R.20

or

ALL RADII R5.

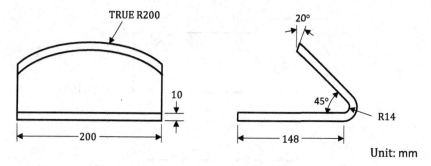

FIGURE 7.36 TRUE R note.

7.3.1.2 Rounded Ends

Overall dimensions should be used for objects or features have that rounded ends:

- For fully rounded ends, the radius R is shown but not dimensioned, as shown in Figure 7.37(a).
- For objects with partially rounded ends, the radius is dimensioned, as shown in Figure 7.37(b).
- When a hole and a radius have the same center and the hole location is more important than the location of a radius, either the radius or the overall length should be presented as reference dimension, as shown in Figure 7.37(c).

7.3.1.3 Chords, Arcs, and Angles

Chords, arcs, and angles are presented differently in graphics. Typical dimensions of specifying chords, arcs, and angles are shown in Figure 7.38.

7.3.1.4 Diameters

When specifying the diameter of a single feature or the diameters of a number of concentric cylindrical features, it is recommended that the dimensions be shown on the longitudinal views. Figure 7.39 shows the typical dimensioning of diameters.

When space is restricted or a view is only partially presented, diameters may be dimensioned, as shown in Figure 7.10. No matter where the diameter dimension is shown, the diameter symbol f in italic is always required prior to the diameter value for both the English system and the SI.

7.3.1.5 Countersinks, Counterdrills, Counterbores, and Spotfaces

- **Countersinks**

  A countersink is an angular-sided recess that accommodates the head of a flathead screw, a rivet, and a similar item. The diameters at the surface and the concluded angle are given. When the depth of the tapered section of the countersink is critical, it is specified in the note or by a dimensioning.

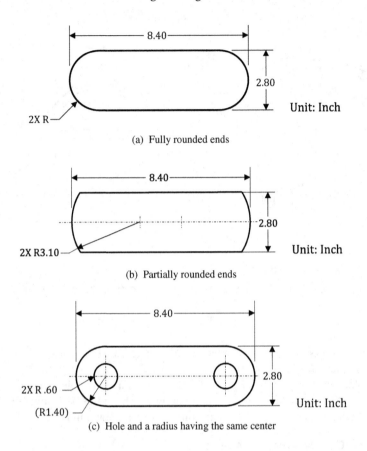

FIGURE 7.37   Dimensioning rounded ends.

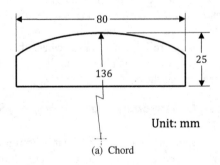

FIGURE 7.38   Chords, arcs, and angles.

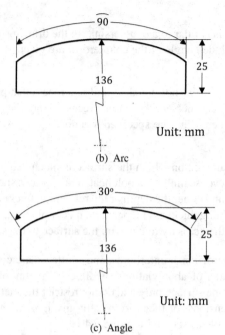

(b) Arc

(c) Angle

FIGURE 7.38   (Continued)

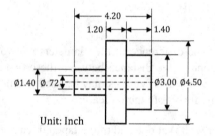

(a) One-view drawing

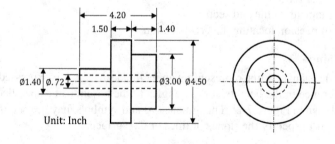

(b) Two-view drawing

FIGURE 7.39   Typical dimensioning diameters.

- **Counterdrills**
  A counterdrill is similar to a countersink. In the drawing, the hole diameter, depth, and included angle of the counterdrill are given.

- **Counterbores**
  A counterbore is a flat-bottomed, cylindrical recess that permits the head of a fastening device, such as a bolt, to lie recessed into the part. The diameter, depth, and corner radius are specified in a note.

- **Spotfaces**
  A spotface is an area on which the surface is machined just enough to provide smooth, level seating for a bolt heat, a nut, or a washer. The diameter of the faced area and either the depth or the remaining thickness are given. If no depth or remaining thickness is specified, it is implied that the spotface is the minimum depth necessary to clean up the surface to the specified diameter.

Countersinks, counterdrills, counterbores, and spotfaces are commonly specified on the drawing by means of abbreviations or dimension symbols. The abbreviations indicate the form of the surface only and do not restrict the methods used to produce that form. The dimensions for them are usually given as a note or preceded by the size of the through hole as shown in Figure 7.40.

7.3.2 COMMON FEATURES

7.3.2.1 Taper

A taper is the ratio of the difference in the diameters of two sections (perpendicular to the axis of a cone to the distance between these two sections) as shown in Figure 7.41. The dimensions and symbols may be used in suitable combinations for specifying the sizes and forms of tapered features:

- The diameter (or width) at one end of the tapered feature
- The length of the tapered feature
- The taper angle
- The taper ratio
- The diameter at the end section
- The dimension locating the cross section

7.3.2.2 Slope

A slope is the slant of a line representing an inclined surface. It is expressed as a ratio of the difference in the heights at right angles to the baseline at a specified distance apart, as shown in Figure 7.42. The dimensions and symbols may be used in different combinations to specify the slopes of lines or flat surfaces:

- The slope specified as a ratio combined with the slope symbol shown in Figure 7.42(a)
- The slope specified by an angle shown in Figure 7.42(b)
- The dimension showing the height from the baseline at the end of the slope shown in Figure 7.42(c)

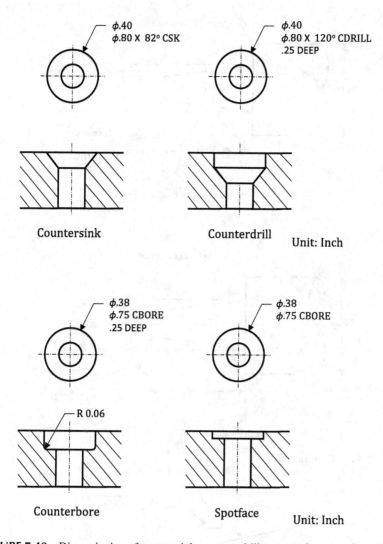

FIGURE 7.40   Dimensioning of countersinks, counterdrills, counterbores, and spotfaces.

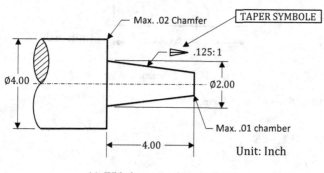

(a) With the taper ratio and symbol

FIGURE 7.41   Dimensioning tapers.

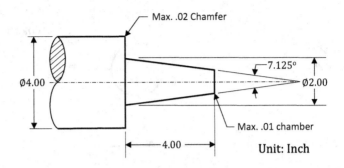

(b) With the taper angle

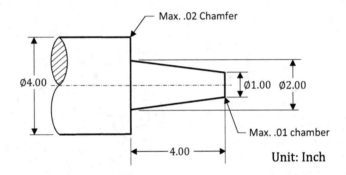

(c) With the diameter at the end section

FIGURE 7.41   (Continued)

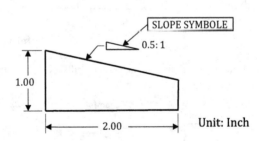

(a) With the slope ratio and symbol

FIGURE 7.42   Dimensioning slopes.

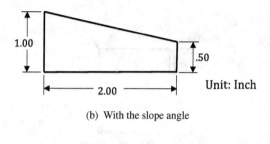

(b) With the slope angle

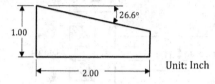

(c) With the height at the end of the slope

FIGURE 7.42 (Continued)

7.3.2.3 Chamfer

A chamfer is a transitional edge between two faces of an object and may some-times regarded as a type of bevel. In engineering applications, chambering is a pro-cess for cutting away the inside or outside piece to facilitate assembly. Chambers are normally dimensioned by giving their angle and linear length, as shown in Figure 7.43.

- When a very small chamfer is permissible, primarily to break a sharp corner, it may need to be dimensioned but not drawn. If not otherwise specified, an angle of 45° is always applied, as shown in Figure 7.43(a).
- Internal chamfers may be dimensioned in the same manner, but giving the diameter over the chamfer is often desirable. The angle may also be given as the included angle if this is a design requirement. This type of dimen-sioning is generally necessary for larger diameters, especially those over 2 in. (50 mm), whereas chamfers on small holes are usually expressed as countersink, as shown in Figure 7.43(b).
- Chamfers are never measured along the angular surface. Chamfers are made between surfaces at other than 90° as shown in Figure 7.43(c).

7.3.2.4 Repetitive Dimension

Dimensioning repetitive features called repetitive dimension may be specified on a drawing by using an "X" or "spaces" in conjunction with the numeral to indicate the "number of times" or "places" as they are required. A space is shown between the X and the dimension. An X that means "by" is often used between coordinate dimensions specified in note form, as shown in Figures 7.44(a), 7.44(b), and 7.44(c).

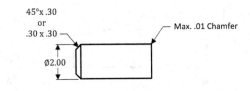

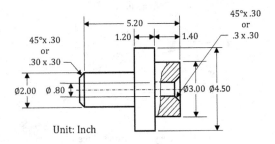

(a) Small and 45° chamfers

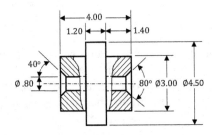

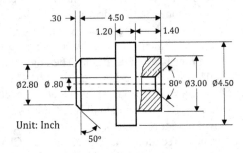

(b) For all chamfers

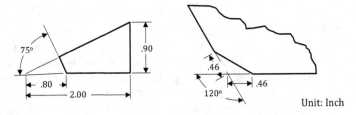

(c) Chamfers between surfaces at other than 90°

FIGURE 7.43 Dimensioning chamfers.

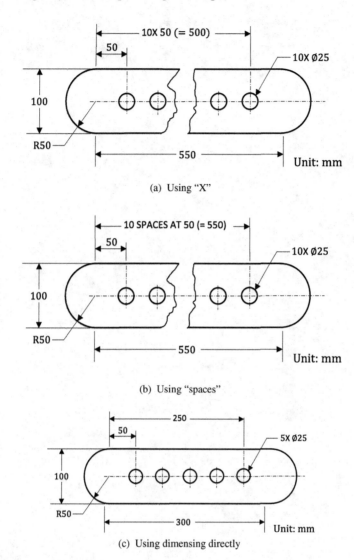

(a) Using "X"

(b) Using "spaces"

(c) Using dimensing directly

FIGURE 7.44 Dimensioning repetitive features.

8 Dimensioning Pipes and Ducts of Pipe and Duct Systems

8.1 INTRODUCTION

8.1.1 PIPE SYSTEMS

In engineering applications, *piping* refers to the pipe system design and the detailed physical pipe system layout within a process plant or commercial building. Pipe systems are commonly used to convey fluids from one location to another location or from one equipment to another equipment. The fluid can be heating water or steam in a heating system, chilled water or refrigerant in a cooling system, cold water or hot water in a plumbing system, compressed air in a compressed air system, and so on. For example, the steam pipe system in a thermal power plant transports steam to the steam turbine generator for generating electricity. In pipe systems, pipes are the basic components. A pipe system is an assembly or a network of pipes with fittings, valves, and other components. The pipe fittings are used to connect pipes to construct a long pipe system. Commonly used fittings are elbows, tees, reducers, and so on. Valves are used to stop, divert, or control the fluid flow. Commonly used valve types are gate valves, globe valves, butterfly valves, ball valves, and control valves. The types of fittings and valves selected in the pipe systems are based on their intended function and application. Pipes applied in the pipe systems may be in a variety of materials, such as steel, stainless, aluminum, brass, copper, glass, or plastic. Commonly used pipe materials are steel and copper. Usually, the pipe fittings and valves are made of the same material as the pipe. In general, the dimension of sectional flow or diameter of the fittings and valves should be the same as the pipes. The purpose of dimensioning on engineering drawings is to provide a clear and complete description of the pipe system design. A complete set of dimensions in the design will permit only one interpretation needed for successful manufacturing and installation. In a design, engineers should ensure that the production team can get all information and data directly from the drawing. Figure 8.1(a) presents a typical pipe system, and Figure 8.1(b) is shows pipe systems in a process plant.

8.1.2 DUCT SYSTEMS

In engineering applications, a duct system is a network of conduits used to deliver and remove gases for heating, ventilation, and air conditioning (HVAC), such as supply air, return air, fresh air and exhaust air. The ventilation duct system is to deliver the fresh air and exchange air. The heating and cooling duct system in HVAC supplies conditioned air to the air-conditioned areas. Like pipe systems, ducts are the basic components in the duct systems. A duct system is an assembly of ducts with

DOI: 10.1201/9781003508977-8

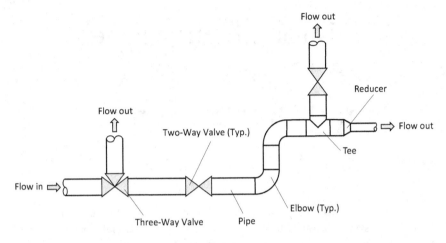

(a) A typical pipe system

(b) Pipe systems in a process plant

FIGURE 8.1 Pipe systems.

fittings, dampers, and other components. The duct fittings are used to connect ducts to construct a large duct system. Commonly used fittings are transitions, takeoffs, and reducers, among others. The dampers are used to stop and control gas flow. Commonly used damper types are opposed blade and parallel blade. In general, dampers are manipulated by actuators. The actuators are powered pneumatically or electrically. The popular material of duct is sheet metal, which typically is galvanized steel or aluminum metal. Figure 8.2(a) presents a typical duct system, and Figure 8.2(b) shows a duct system in a building.

In duct systems, flexible ducts, called flexible hoses, are commonly applied to connect from ducts to the end users. As the name implies, the flexible ducts are not rigid.

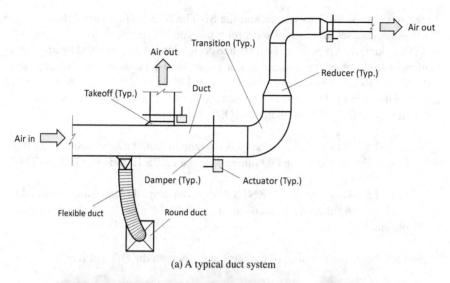

(a) A typical duct system

(b) A duct system in a building

FIGURE 8.2 Duct system.

They are made with steel spring coils covered by thick plastics. The major advantage of using the flexible duct is that the connection is flexible and can be maneuvered easily, for example, when the diffuser location is uncertain in design or may be altered later.

8.2 PIPING DIMENSIONS

8.2.1 PIPE SIZES

Commonly used metallic pipes in engineering practice are steel and copper. The sizes of pipes, fittings, and valves are usually specified in inches by the nominal pipe size (NPS) or in millimeters by the diameter nominal (DN) to denote the diameter dimensions in the English system and the SI. The NPS and DN are defined by the North American set of standard sizes for pipes and ISO 6708. When using the NPS and DN to identify a pipe size, the pipe size designation is not followed by any abbreviation of the unit of measurement. For example, a pipe diameter will appear as NPS 1/2 or DN 25, which is not NPS 1/2 in. or DN 25 mm. The designations of 1/2 and 25 right after the NPS and DN are not the physical dimensions, neither the internal diameter (ID) nor the out diameter (OD) of the pipe.

- For NPS 1/8 to NPS 12—The NPS designation and OD dimension are different. For example, the OD dimension of an NPS 12 pipe is 12.75 in (324 mm) physically.
- For NPS 14 and above—The NPS designation and OD dimension are equal. For example, the OD dimension of an NPS 14 pipe is 14 in (360 mm) physically.

Figure 8.3 shows the relations of the NPS and DN with the OD and ID.

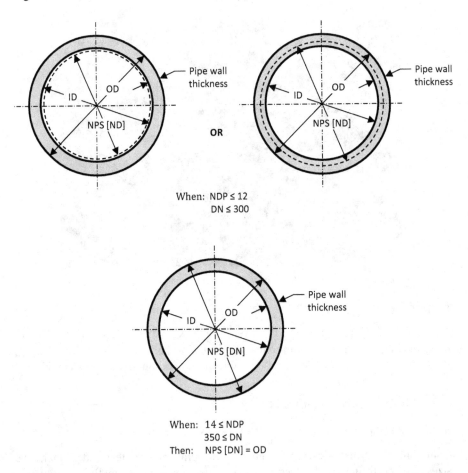

FIGURE 8.3 Relations of the NPS and DN with the OD and ID.

In engineering practice, to find the physical OD and ID dimension corresponding to the NPS or DN designations, a table, such as Table 8.1, is commonly used. The table is excerpted from the ASME standards B36.10M and B36.19M.

TABLE 8.1
OD and ID of NPS and DN Schedule 40 Steel Pipe

| Pipe Size NPS (in) | Diameter | | Nominal | Weight | |
DN (mm)	Internal	External	Thickness	lb/ft	kg/m
1/8	0.405	0.27	0.07	0.24	0.36
6	10.3	6.86	1.78		
1/4	0.54	0.36	0.09	0.42	0.63
8	13.7	9.14	2.29		
3/8	0.675	0.49	0.09	0.57	0.84
10	17.1	12.4	2.29		
1/2	0.84	0.62	0.11	0.85	1.26
15	21.3	15.7	2.79		
3/4	1.05	0.82	0.11	1.13	1.68
20	26.7	20.8	2.79		
1	1.315	1.05	0.13	1.68	2.5
25	33.4	26.7	3.3		
1¼	1.66	1.38	0.14	2.27	3.38
32	42.2	35.1	3.56		
1½	1.9	1.61	0.15	2.72	4.04
40	48.3	40.9	3.81		
2	2.375	2.07	0.15	3.65	5.43
50	60.3	52.6	3.81		
2½	2.875	2.47	0.2	5.79	8.62
65	73	62.7	5.08		
3	3.5	3.07	0.22	7.58	11.27
80	88.9	78	5.59		
4	4.5	4.03	0.24	10.79	16.06
100	114	102	6.1		
5	5.563	5.05	0.26	14.61	21.74
125	141	128	6.6		
6	6.625	6.07	0.28	18.97	28.23
150	168	154	7.11		
8	8.625	7.98	0.32	28.55	42.49
200	219	203	8.13		
10	10.75	10.02	0.37	40.48	60.24
250	273	255	9.4		
12	12.75	11.94	0.41	53.6	79.77
300	324	303	10.4		
14	14	13.13	0.44		93.75
350	356	334	11.2	63	

Steel

Commercial steel pipes are made in standard sizes. For a given pipe NPS or DN, the pipe can have several different wall thicknesses that depend on the pipe schedules. The bigger the schedule number is, the larger the wall thickness is.

- For a given NPS or DN, when the OD stays fixed, the wall thickness increases with increase in schedule number. In other words, the actual flow sectional area reduces.
- For a given schedule number, when the OD increases with the NPS, then the wall thickness either stays constant or increases.

For example, the wall thickness of an NPS 14 schedule 60 steel pipe is thicker than that of an NPS 14 schedule 40 steel pipe, which can be seen from the following table.

NPS	DN	Outside Diameter		Wall Thickness			
				SCH 40		SCH 60	
in.	mm	in.	mm	i.n	mm	in.	mm
14	350	14.00	355.60	0.44	11.13	0.594	15.09

In engineering applications, the most important pipe information is the pipe ID. The ID of a pipe can be calculated conveniently once pipe NPS (or DN) and pipe schedule are decided. The pipe ID is determined by the pipe OD of the pipe NPS (or DN) minus double the pipe wall thickness.

Exercise 8.1

An NPS 12 schedule 40 steel pipe is specified. Find the pipe's (a) OD, (b) ID, and (c) wall thickness in inches and millimeters, respectively.

SOLUTION

Referring to Table 8.1 and using NPS 12 schedule 40 steel pipe, the pipe is DN 300 in the SI,

- (a) NPS 12: OD is 12.75 in.;
     DN 300: OD is 324 mm.
- (b) NPS 12: ID is 11.94 in.;
     DN 300: ID is 303 mm.
- (c) NPS 12: wall thickness is 0.41 in.;
     DN 300: wall thickness is 10.4 mm.

Alternatively, using the relation of OD, wall thickness, and ID to calculate ID,

$$\text{pipe OD} - 2 \times \text{wall thickness} = \text{pipe ID}$$

then,

NPS 12: pipe ID = 12.75 in. $- 2 \times 0.41$ in. $\doteq 11.94$ in.

DN 300: pipe ID = 324 mm $- 2 \times 10.4$ mm $\doteq 303$ mm

Or finding the wall thickness by using

$$\frac{\text{pipe OD} - \text{pipe ID}}{2} = \text{wall thickness}$$

so

NPS 12: $\text{wall thickness} = \dfrac{12.75 \text{ in} - 11.94 \text{ in}}{2} \doteq 0.41 \text{in}$

DN 300: $\text{wall thickness} = \dfrac{324 \text{ mm} - 303 \text{ mm}}{2} \doteq 10.4 \text{mm}$

Copper

The standard copper pipe size is often measured in the NPS or DN. Commonly used copper pipe sizes in residential plumbing are NPS 1/2 (DN 15), NPS3/4 (DN 20), and NPS 1 (DN 25). Commercial and industrial applications may use larger sizes, such as NPS 2 (DN 50) and NPS 3 (DN 80). Commonly used copper tubes are classified as Type K, Type L, and Type M corresponding to different wall thicknesses as per the ASTM standard B88 and B306. The wall thickness and working pressure are reduced from Type K through Type M. For example, an NPS 2 Type K copper pipe has an OD of 2.125 in. (53.975 mm) and a wall thickness of 0.083 in. (2.108 mm), while the wall thicknesses are 0.070 in. (1.778 mm) and 0.058 in. (1.473 mm) for Type L and Type M copper pipes, respectively. Table 8.2 shows the copper pipe dimensions for plumbing application.

Wall thicknesses used for plumbing in the United States and Canada are the CTS—copper tube sizes—Type K, L and M.

Exercise 8.2 (FE style)

Select the right internal diameter ID and OD of an NPS 1 Type L copper pipe.

(A)   1.025, 1 1/8
(B)   26.036, 28.575
(C)   1 1/8, 1.025
(D)   28.575, 26.036

SOLUTION

The correct answers are **(A)** and **(B)**. (A) is in inches, and (B) is in mm. (C) and (D) are wrong since the answers are in the order of ID and OD.

TABLE 8.2

Copper Pipe Sizes in Plumbing

Nominal Pipe Size (NPS)	Outside Diameter (OD)		Inside Diameter (ID)					
			Type K		Type L		Type M	
	in	mm	in	mm	in	mm	in	mm
1/4	3/8	9.5	0.305	7.747	0.315	8.001		
3/8	1/2	12.7	0.402	10.211	0.43	10.922	0.45	11.43
1/2	5/8	15.875	0.528	13.411	0.545	13.843	0.569	14.453
5/8	3/4	19.05	0.652	16.561	0.668	16.967	0.69	17.526
3/4	7/8	22.225	0.745	18.923	0.785	19.939	0.811	20.599
1	1 1/8	28.575	0.995	25.273	1.025	26.035	1.055	26.797
11/4	1 3/8	34.925	1.245	31.623	1.265	32.131	1.291	32.791
11/2	1 5/8	41.275	1.481	37.617	1.505	38.227	1.527	38.786
2	2 1/8	53.975	1.959	49.759	1.985	50.419	2.009	51.029
21/2	2 5/8	66.675	2.435	61.849	2.465	62.611	2.495	63.373
3	3 1/8	79.375	2.907	73.838	2.945	74.803	2.981	75.717

8.2.2 DIMENSIONING PIPE SYSTEMS

Steel pipes are commonly joined by methods of welded, flanged, threaded, or grooved-end fittings. Hybrid joints are common in a piping system, such as typically using threaded fittings for pipes NPS 2 (DN 50) and below and flanged fittings for sizes NPS 2.5 (DN 65) and above. While copper pipes are commonly joined by methods of soldering, brazing, and press fittings. Brazing is that is for the dimension of NPS 2 (DN 50) and above and press fittings are popular for the dimension below NPS 2 (DN 50). Figure 8.4 shows a drawing plane of pipe systems in a process plant and Figure 8.5 presents a typical side view of pipe systems showing elevation dimensions, respectively. In the figures, the pipe dimensions are NPS. For example, 2″ steel means NPS 2 steel pipe.

8.3 DUCTWORK DIMENSIONS

8.3.1 DUCT SIZES

Ductwork refers to a system of ducts in a network of conduits. A duct can come in various shapes, such as round, rectangular, and oval. The choice of duct size involves the airflow rate Q and the air velocity V. The flow cross-sectional area A of a duct can be found by using the formula of

$$A = \frac{Q}{V}$$

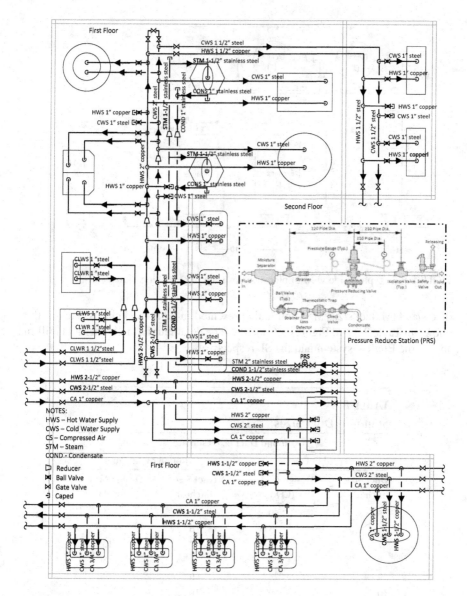

FIGURE 8.4 A drawing plane of pipe systems in a process plant (Schedule 40 steel and Type L copper).

for either a round duct or rectangle duct. In engineering practice, standards for duct sizes are established by the organizations such as the American Society of Heating, Refrigerating and Air-Conditioning Engineers (ASHRAE). These standards are based on extensive research and experiments that consider many factors, for example, airflow velocity, pressure loss, noise level, and energy efficiency, among others.

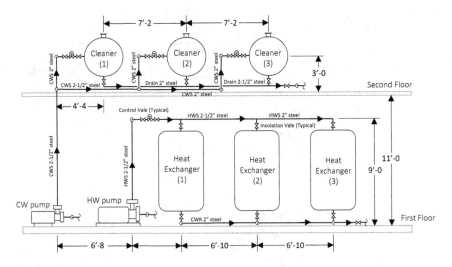

FIGURE 8.5 Typical side view of pipe systems showing elevation dimensions.

The standard duct sizes are commonly presented in tables and charts, such as those shown in Table 8.3 and Figures 8.6 and 8.7. The default units are inches and millimeters in the English system and the SI, respectively.

TABLE 8.3
Standard Duct Sizes

	Round Duct Size Estimate		
Flexible Duct		Round Metal Duct	
Duct Size (inches)	Design Airflow (CFM)	Duct Size (inches)	Design Airflow (CFM)
5	50	5	50
6	75	6	85
7	110	7	125
8	160	8	180
9	225	9	240
10	300	10	325
12	480	12	525
14	700	14	750
16	1,000	16	1,200
18	1,300	18	1,500
20	1,700	20	2,000

Note: Friction resistance = typically 0.05" per 100' in sheet metal flexible duct and 0.06" per 100' typically in sheet metal round duct.

Rectangular Duct Size Estimate									
Design Airflow	Duct Height (net inside dimension, inches)								
(CFM)	4	(CFM)	6	(CFM)	8	(CFM)	10	(CFM)	12
60	6×4	60	4×6	90	4×8	120	4×10	150	4×12
90	8×4	110	6×6	160	6×8	215	6×10	270	6×12
120	10×4	160	8×6	230	8×8	310	8×10	400	8×12
150	12×4	215	10×6	310	10×8	430	10×10	550	10×12
180	14×4	270	12×6	400	12×8	550	12×10	680	12×12
210	16×4	320	14×6	490	14×8	670	14×10	800	14×12
240	18×4	375	16×6	580	16×8	800	16×10	950	16×12
270	20×4	430	18×6	670	18×8	930	18×10	1,100	18×12
300	22×4	490	20×6	750	20×8	1,060	20×10	1,250	20×12
330	24×2	540	22×6	840	22×8	1,200	22×10	1,400	22×12
		600	24×6	930	24×8	1,320	24×10	1,600	24×12
		650	26×6	1,020	26×8	1,430	26×10	1,750	26×12
		710	28×6	1,100	28×8	1,550	28×10	1,950	28×12
		775	30×6	1,200	30×8	1,670	30×10	2,150	30×12
				1,300	32×8	1,800	32×10	2,300	32×12
				1,400	34×8	1,930	34×10	2,450	34×12
				1,500	36×8	2.060	36×10	2,600	36×12
						2,200	38×10	2,750	38×12
						2,350	40×10	2,900	40×12
								3,050	42×12

Note:

Friction resistance = typically 0.07" per 100' typically in sheet metal duct.

Instructions for using the table:

1st step—Identify the rate of the airflow in the duct.

2nd step—Select the duct size from the table corresponding to the rate of the airflow.

3rd step—If desired airflow exceeds the design CFM rating, increase to the next duct size.

4th step—The listed CFMs are based on typical field results and may vary per the installation condition, such as dampers.

5th step—If duct run exceeds 25" or has excessive transitions, increase to the next size.

6th step—The duct design alone might be inadequate. The design may need to be proved by test and balanced.

One of useful tools for dimensioning ducts called the duct calculator is popularly used in engineering practice as shown in Figure 8.8. The tool is convenient to use. Once an airflow rate and an air pressure drop in inches of H_2O column per 100 ft in the duct are decided, the corresponding air velocity in the duct, rectangular duct dimension, and equivalent round duct diameter can be obtained quickly from the duct calculator.

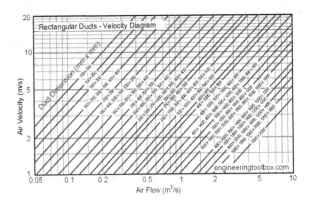

(a) Rectangular duct dimensions in the SI

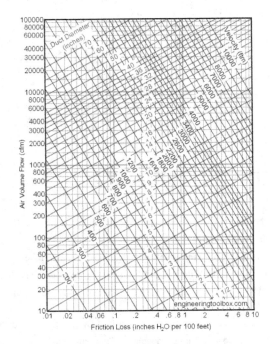

(b) Round duct dimensions in the English system

FIGURE 8.6 Rigid duct dimensions.

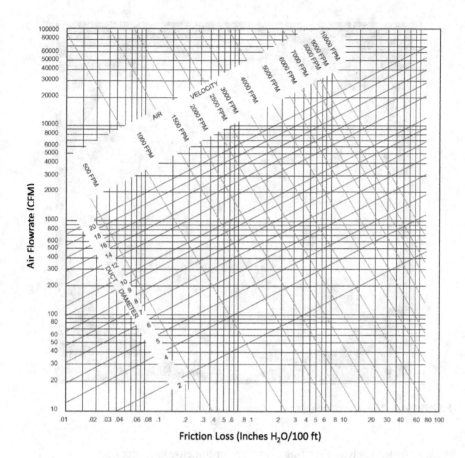

FIGURE 8.7 Flexible duct dimensions.

FIGURE 8.8 The duct calculator.

Some rules apply to selecting duct sizes or dimensioning ducts:

- Standard rectangular duct sizes typically are even numbers and incremented in even numbers.
- On drawings, the format dimensioning ducts is
  Round ØXX, for example, Ø20.
  Rectangular XXxXX; the first value is the size of the duct side facing to the viewer, and the second value is the size of the duct side parallelly with the view of the viewer.
- In all possible, the rectangular duct is preferred to be used in the duct systems if without any reasons for using other duct shapers.

For example, knowing the required air flow rate in a rectangular duct is 400 cfm, the preferred duct size is 12 in. by 8 in. On the drawing plane, the duct dimension is marked as 12×8.

Exercise 8.3 (FE style)

Knowing a duct conveys an air flow 1,500 CFM at the velocity of 1,000 FPM in the duct, select all applicable duct sizes shown.

(A) 22×12
(B) 12×22
(C) 16×12
(D) Ø18

SOLUTION

The correct answers are **(A)** and **(D)**. (B) shows a wrong order. Using the equation, the required flow sectional area is

$$A = \frac{Q}{V}$$

So,

$$A = \frac{1{,}500\,\dfrac{ft^3}{min}}{1{,}000\,\dfrac{ft}{min}} = 1.5\,ft^2$$

Using unit conversion factor,

$$1\ ft = 12\ in.$$

in the previous expression, the area becomes

$$A = 1.5 \times 12^2\ in.^2 = 216\ in.^2$$

Therefore, (C) the flow sectional area is not enough. The rectangular 22×12 and round duct Ø18 meet the dimension of the flow sectional area, that is,

$$22\ in. \times 12\ in. = 264\ in.^2$$

and

$$\frac{\pi}{4}\,(18\,in.)^2 = 255\ in.^2$$

Exercise 8.4

The ductwork from the HVAC system to the diffuser is shown in Figure *Exercise 8.4*. Knowing that the friction loss calculated in the assembly is 0.08 in H_2O/100 ft and the airflow rate is 1,000 CFM, determine (a) the diameter of the round duct, (b) the diameter of the flexible duct, (c) the size of the rectangular duct, and (d) the air velocity in the assembly.

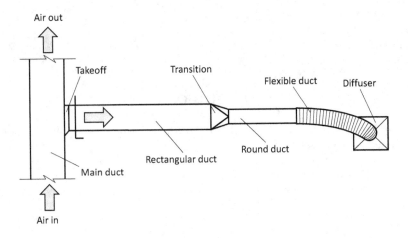

FIGURE *Exercise 8.4.*

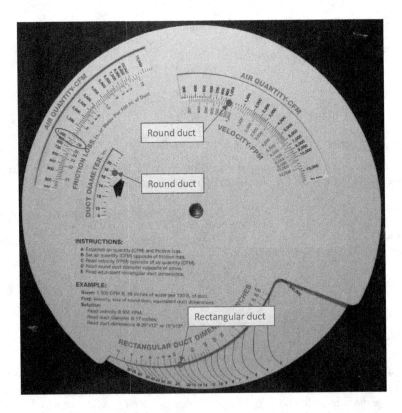

FIGURE Solution for *Exercise 8.4.*

SOLUTION

(a) Referring to Table 8.1 or Figure 8.4, the diameter of the round duct for the airflow rate 1,000 CFM is **16 in.**
(b) Referring to Table 8.1 or Figure 8.5, the diameter of the flexible duct for the airflow rate 1,000 CFM is **16 in.**
(c) Referring to Table 8.1 or the duct calculator as shown, the size of the rectangular duct can be obtained as **18 in. × 12 in.**
(d) Referring to the duct calculator as shown, the air velocity in the assembly is **890 feet per minute.**

Exercise 8.5 (FE style)

Knowing that the airflow rate is 930 CFM and the air pressure drop is 0.08 in H_2O column per 100 ft in the duct, select all applicable duct sizes shown.

(A) 24 in. × 8 in.
(B) 18 in. × 10 in.
(C) 16 in. × 12 in.
(D) 12 in. × 16 in.

SOLUTION

The correct answers are **(A)**, **(B)**, **(C)**, and **(D)**. By using the duct calculator, the result can be quickly obtained. The difference between (C) and (D) is that (C) may be 16 in. in width and 12 in. in height, and (D) is just opposite.

8.3.2 DIMENSIONING DUCT SYSTEMS

As mentioned in Section 8.2.2, it is a common practice to use flexible ducts connecting from the ducts to the end users, such as the diffusers, grilles, vents, registers, and louvers. Figure 8.9 shows a drawing plane of HVAC duct systems in a building, and Figure 8.10 presents a side view of the duct systems showing elevation dimensions, respectively.

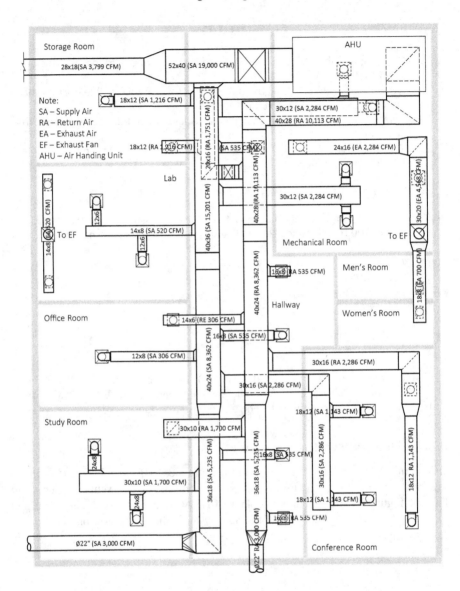

FIGURE 8.9 A drawing plane of HVAC duct systems in a building.

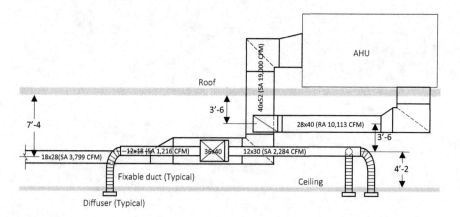

FIGURE 8.10 A side view of the duct systems showing elevation dimensions.

Appendix

A.1 THE INTERNATIONAL SYSTEM OF UNITS

The International System of Units (SI) is developed and published coordinately by three organizations:

- The International Bureau of Weights and Measures (BIPM)
- The International Committee for Weights and Measures (CIPM)
- The General Conference on Weights and Measures (CGPM)

The BIPM is the organization founded on May 20, 1875, the CGPM is the general meeting of the shareholders, and the CIPM is the board of directors appointed by the CGPM. The BIPM operates under the exclusive supervision of the CIPM, which itself comes under the authority of the CGPM and reports to it on the work accomplished by the BIPM.

Name	International Bureau of Weights and Measures (BIPM)	International Committee for Weights and Measures (CIPM)	General Conference on Weights and Measures (CGPM)
In French	Bureau International des Poids et Mesures	Comité international des poids et mesures	Conférence générale des poids et mesures
Function	• represent the worldwide measurement community, aiming to maximize its uptake and impact • be a center for scientific and technical collaboration between Member States, providing capabilities for international measurement comparisons on a shared-cost basis • be the coordinator of the worldwide measurement system, ensuring it gives comparable and internationally accepted measurement results	• discuss the work of the BIPM under the delegated authority of the CGPM • issue an annual report on the administrative and financial position of the BIPM to the governments of the state parties to the Meter Convention • discuss metrological work that member states decide to do in common, and set up and coordinate activities between specialists in metrology • make appropriate recommendations • commission reports in preparation for meetings of the CGPM	• discuss and initiate the arrangements required to ensure the propagation and improvement of the International System of Units (SI), which is the modern form of the metric system • confirm the results of new fundamental metrological determinations and various scientific resolutions of international scope • take all major decisions concerning the finance, organization, and development of the BIPM
Meeting		Every year (in two sessions per year)	Every four years at the BIPM

(Continued)

(Continued)

Name	International Bureau of Weights and Measures (BIPM)	International Committee for Weights and Measures (CIPM)	General Conference on Weights and Measures (CGPM)
Address	Postal address: Pavillon de Breteuil, F-92312 Sèvres Cedex. Paris, France Street address: 12bis Grande Rue, F-92310 Sèvres	Saint-Cloud, Hauts-de-Seine, France	

CGPM

There are 64 members currently in the CGPM. Each member represents a different country.

No.	Country	Year Jointed	No.	Country	Year Jointed
Current Members (until 2022)					
1	Argentina	1877	33	Kenya	2010
2	Australia	1947	34	Lithuania	2015
3	Austria[1]	1875	35	Malaysia	2001
4	Belarus	2020	36	Mexico	1890
5	Belgium	1875	37	Montenegro	2018
6	Brazil	1921	38	Morocco	2019
7	Bulgaria	1911	39	Netherlands	1929
8	Canada	1907	40	New Zealand	1991
9	Chile	1908	41	Norway[4]	1875
10	China	1977	42	Pakistan	1973
11	Colombia	2012	43	Poland	1925
12	Costa Rica	2022	44	Portugal	1876
13	Croatia	2008	45	Romania	1884
14	Czech Republic[2]	1922	46	Russia[5]	1875
15	Denmark	1875	47	Saudi Arabia	2011
16	Ecuador	2019	48	Serbia	2001
17	Egypt	1962	49	Singapore	1994
18	Estonia	2021	50	Slovakia[2]	1922
19	Finland	1913	51	Slovenia	2016
20	France	1875	52	South Africa	1964
21	Germany[3]	1875	53	South Korea	1959
22	Greece	2001	54	Spain	1875

(Continued)

No.	Country	Year Jointed	No.	Country	Year Jointed
23	Hungary[1]	1925	55	Sweden[4]	1875
24	India	1880	56	Switzerland	1875
25	Indonesia	1960	57	Thailand	1912
26	Iran	1975	58	Tunisia	2012
27	Iraq	2013	59	Turkey[6]	1875
28	Ireland	1925	60	Ukraine	2018
29	Israel	1985	61	United Arab Emirates	2015
30	Italy	1875	62	United Kingdom	1884
31	Japan	1885	63	United States	1878
32	Kazakhstan	2008	64	Uruguay	1908

Former Members (until 2022)

No.	Country	Period	No.	Country	Period
1	Cameroon	1970–2012	4	Peru	1875–1956
2	Dominican Republic	1954–2015	5	Venezuela	1879–1907
					1960–2018
3	North Korea	1982–2012			

[1] Joined originally as Austria-Hungary.

[2] Joined originally as part of Czechoslovakia.

[3]. Joined originally as the German Empire.

[4] Joined originally as part of Sweden and Norway.

[5] Joined originally as the Russian Empire.

[6] Joined originally as the Ottoman Empire.

There are 36 associates currently in the CGPM. Each associate represents a different country. The associates for the countries not yet as members and for economic unions.

No.	Country	Year Jointed	No.	Country	Year Jointed
Current Associates (until 2022)					
1	Albania	2007	19	Mauritius	2010
2	Azerbaijan	2015	20	Moldova	2007
3	Bangladesh	2010	21	Mongolia	2013
4	Bolivia	2008	22	Namibia	2012
5	Bosnia and Herzegovina	2011	23	North Macedonia	2006
6	Botswana	2012	24	Oman	2012
7	Cambodia	2021	25	Panama	2003
8	Caribbean Community	2005	26	Paraguay	2009
9	Chinese Taipei	2002	27	Peru	2009
10	Ethiopia	2018	28	Philippines	2002
11	Georgia	2008	29	Qatar	2016

(Continued)

(Continued)

No.	Country	Year Jointed	No.	Country	Year Jointed
12	Ghana	2009	30	Sri Lanka	2007
13	Hong Kong	2000	31	Syria	2012
14	Jamaica	2003	32	Tanzania	2018
15	Kuwait	2018	33	Uzbekistan	2018
16	Latvia	2001	34	Vietnam	2003
17	Luxembourg	2014	35	Zambia	2010
18	Malta	2001	36	Zimbabwe	2010–2020 2022

Former Associates

No.	Country	Period	No.	Country	Period
1	Cuba	2000–2021	3	Sudan	2014–2021
2	Seychelles	2010–2021			

CIPM

There are 18 members currently in the CIPM. Each member represents a different country. There are 10 Consultative Committees currently, as shown in the following table, whose function is to provide information on matters referred to them for study and advice.

No.	Abbreviation	Consultative Committee	Year Founded
1	CCEM	For Electricity and Magnetism	1927
2	CCPR	For Photometry and Radiometry	1933
3	CCT	for Thermometry	1937
4	CCL	For Length.	1952
5	CCTF	For Time and Frequency	1956
6	CCRI	For Ionizing Radiation	1958
7	CCU	For Units	1954
8	CCM	For Mass and Related Quantities	1980
9	CCQM	For Amount of Substance: Metrology in Chemistry and Biology	1993
10	CCAUV	For Acoustics, Ultrasound and Vibration	1999

BIPM

There are 45 physicists and technicians work in BIPM laboratories. They mainly conduct metrological research, international comparisons of realizations of units, and the calibrations of standards. An annual director's report gives details of the work in progress.

A.2   QUANTITIES, DIMENSIONS, AND UNITS OF THE SI AND THE ENGLISH SYSTEM

(a) Fundamental:

	Dimension	SI Unit		English Unit	
Quantity		Name	Symbol	Name	Symbol
Mass m	M	kilogram	kg	pound	lbm
Length l	L	meter	m	foot	ft
Time t	t (θ or T)	second	s	second	s
Thermodynamic temperature T	T	Kelvin	K	Rankine	R
Electric current i	I	ampere	A	ampere	A
Luminous intensity Iv	J	candela	cd	candela	cd
Amount of substance n	N	mole	mol	mole	mol

Note: If both temperature and time are presented in an application, the dimension symbol of temperature is T, and time can be either t or θ.

(b) Derived:

	SI Unit			English Unit		
	Name	Symbol	Derived in Fundamental Unit	Name	Symbol	Derived in Fundamental Unit
Quantity						
Acceleration a	meter per square second	$m \cdot s^{-2}$	m/s^2	feet per square second	$ft \cdot s^{-2}$	ft/s^2
Area A	square meter	m^2	m^2	square foot	ft^2	ft^2
Angular acceleration ω	radian per second	$Rad.s^{-1}$	$1/s$	radian per second	$Rad.s^{-1}$	$1/s$
Angular velocity α	radian per square second	$Rad.s^{-2}$	$1/s^2$	radian per square second	$Rad.s^{-2}$	$1/s^2$
Coefficient of thermal expansion β	the inverse of temperature	K^{-1}	$1/K$	the inverse of temperature	R^{-1}	$1/R$
Density ρ	mass per cubic meter	$kg \cdot m^{-3}$	kg/m^3	mass per cubic foot	$lbm \cdot ft^{-3}$	lbm/ft^3
Electric charge density ρ	Coulomb per cubic meter	$C \cdot m^{-3}$	$A \cdot s/m^3$	Coulomb per cubic foot	$C \cdot ft^{3}$	$A \cdot s/ft^3$
Electric field strength E	volt per meter	$V \cdot m^{-1}$	$kg \cdot m/s^3 \cdot A$	volt per foot	$V \cdot ft^{-1}$	$lbm \cdot ft/s^3 \cdot A$
Energy E	joule	J $N \cdot m$	$kg \cdot m^2/s^2$	foot pound-force	$ft \cdot lbf$	$lbm \cdot ft^2/s^2$

(*Continued*)

(Continued)

Quantity	SI Unit			English Unit		
	Name	Symbol	Derived in Fundamental Unit	Name	Symbol	Derived in Fundamental Unit
Enthalpy H	joule	kJ	$kg.m^2/s^2$	British thermal unit	Btu	$kg.m^2/s^2$
Frequency f	hertz	Hz	$1/s$	hertz	Hz	$1/s$
Force F	newton	N	$kg.m/s^2$	pound-force	lbf	$lbm.ft/s^2$
Friction f	newton	N	$kg.m/s^2$	pound-force	lbf	$lbm.ft/s^2$
Gas constant R	–	$kJ \cdot kg^{-1} \cdot K^{-1}$	$m^2/s^2 \cdot K$	–	$Btu.lbm^{-1}.R^{-1}$	$ft^2/s^2 \cdot R$
Gravitational acceleration g	meter per square second	$m \cdot s^{-2}$	m/s^2	foot per square second	$ft \cdot s^{-2}$	ft/s^2
Gravitational constant g_c	–	–	–	–	$lbm.ft \cdot lbf^{-1} \cdot s^{-2}$	1
Heat Q	joule	J	$kg \cdot m^2/s^2$	British thermal unit	Btu	$lbm \cdot ft^2/s^2$
Heat capacity C	joule per kelvin	$J \cdot K^{-1}$	$kg \cdot m^2/s^2 \cdot K$	Btu per rankine	$But \cdot R^{-1}$	$blm \cdot ft^2/s^2 \cdot R$
Heat flow rate $\dot{Q}$	heat flow per unit of time	$J \cdot s^{-1}$ W	$kg \cdot m^2/s^3$	Heat flow per unit of time	$Btu \cdot h^{-1}$	$lbm \cdot ft^2/s^3$
Heat transfer coefficient of convection h	heat flowrate per square meter kelvin	$W \cdot m^{-2} \cdot K^{-1}$	$kg/s^3 \cdot K^{-1}$	heat flowrate per square foot rankine	$Btu \cdot h^{-1} \cdot ft^{-2} \cdot R^{-1}$	$lbm/s^3 \cdot R^{-1}$
Height z	meter	h	m	feet	h	ft
Internal Energy U	joule	kJ	$kg \cdot m^2/s^2$	British thermal unit	Btu	$lbm \cdot ft^2/s^2$
Irradiance E	watt per square meter	$W \cdot m^{-2}$	kg/s^3	Btu per second and square foot	$Btu.s^{-1}ft^{-2}$	lbm/s^3
Kinematic viscosity v	ratio of the viscosity to the density	$m^2 \cdot s^{-1}$	m^2/s	ratio of the viscosity to the density	$ft^2 \cdot s^{-1}$	ft^2/s
Mass flow rate	mass flow per second	$kg \cdot s^{-1}$	kg/s	mass flow per second	$lbm \cdot s^{-1}$	lbm/s
Mechanical equivalent heat J	–	–	–	Joule's constant	$ft \cdot lb \cdot Btu^{-1}$	$ft \cdot lb/Btu$
Moment M	newton meter	$N \cdot m$	$kg \cdot m^2/s^2$	foot pound-force	$lbf \cdot ft$	$lbm \cdot ft^2/s^2$
Molar energy U_e	Joule per mole	$J \cdot mol^{-1}$	$kg \, m^2 \, s^{-2} \, mol^{-1}$	Btu per mole	$Btu \cdot mol^{-1}$	$lbm \, ft^2 \, s^{-2} \, mol^{-1}$

(Continued)

Quantity	SI Unit			English Unit		
	Name	Symbol	Derived in Fundamental Unit	Name	Symbol	Derived in Fundamental Unit
Molar mass M	mass of one mole	$kg \cdot mol^{-1}$	kg/mol	mass of one mole	$lbm \cdot mol^{-1}$	lbm/mol
Overall Heat transfer coefficient U	heat flowrate per square meter kelvin	$W \cdot m^{-2} \cdot K^{-1}$	$kg/s^3 \cdot K$	heat flowrate per square foot rankine	$Btu \cdot s^{-1} \cdot m^{-2} \cdot R^{-1}$	$lbm/s^3 \cdot R$
Power	watt	W $J \cdot s^{-1}$	$kg \cdot m^2/s^3$	horsepower	hp $lbf \cdot ft \cdot s^{-1}$	$lbm \cdot ft^2/s^3$
Pressure P	pascal	Pa	N/m^2	–	psi	lbf/in^2
Radiance L	watt per square meter steradian	$W \cdot sr^{-1} \cdot m^{-2}$	kg/s^3	Btu per second square meter steradian	$Btu \cdot s^{-1} \cdot sr^{-1} \cdot m^{-2}$	lbm/s^3
Radiant intensity I_e	watt per steradian	$W \cdot sr^{-1}$	$kg \cdot m^2/s^3$	Btu per second steradian	$Btu \cdot s^{-1} \cdot sr^{-1}$	$lbm \cdot ft^2/s^3$
Specifi energy	Joule per kilogram	$J \cdot kg^{-1}$	m^2/s^2	Btu per pound-mass	$Btu \cdot lbm^{-1}$	ft^2/s^2
Specific heat at constant Pressure c_p	Joule per kilogram kelvin	$J \cdot K^{-1} \cdot kg^{-1}$	$m^2/s^2 \cdot K$	Btu per pound-mass rankine	$Btu \cdot R^{-1} \cdot lbm^{-1}$	$ft^2/s^2 \cdot R$
Specific heat at constant volume c_v	Joule per kilogram kelvin	$JK^{-1} \cdot kg^{-1}$	$m^2/s^2 \cdot K$	Btu per pound-mass rankine	$Btu \cdot R^{-1} \cdot lbm^{-1}$	$ft^2/s^2 \cdot R$
Specific internal energy u	Joule per kilogram	$J \cdot kg^{-1}$	m^2/s^2	Btu per pound-mass	$Btu \cdot lbm^{-1}$	ft^2/s^2
Specific volume v	volume per kilogram	$m^3 \cdot kg^{-1}$	m^3/kg	volume per pound-mass	$ft^3 \cdot lbm^{-1}$	ft^3/lbm
Specific weight γ	weight per cubic meter	$N \cdot m^{-3}$	$kg/s^2 \cdot m^2$	weight per cubic foot	$Lbf \cdot ft^{-3}$	$lbm/s^2 ft^2$
Specific enthalpy h	Joule per kilogram	$J \cdot kg^{-1}$	m^2/s^2	Btu per pound-mass	$Btu \cdot lbm^{-1}$	ft^2/s^2
Surface tension σ	newton per meter	$N \cdot m^{-1}$	kg/s^2	pound-force per foot	$Lbf \cdot ft^{-1}$	lbm/s^2
Thermal conductivity k	watt per meter kelvin	$W \cdot m^{-1} \cdot K^{-1}$	$kg \cdot m/s^3 \cdot K$	Btu per second foot rankine	$Btu \cdot s^{-1} \cdot m^{-1} \cdot R^{-1}$	$lbm \cdot ft/s^3 \cdot R$
Thermal diffusivity α	ratio of thermal conductivity to the product of density and specific heat capacity	$m^2 \cdot s^{-1}$	m^2/s	ratio of thermal conductivity to the product of density and specific heat capacity	$ft^2 \cdot s^{-1}$	ft^2/s

(Continued)

(*Continued*)

Quantity	SI Unit			English Unit		
	Name	Symbol	Derived in Fundamental Unit	Name	Symbol	Derived in Fundamental Unit
Torque τ	newton meter	N.m	kg.m²/s²	foot pound-force	lbf.ft	lbm.ft²/s²
Universal gas constant R_u	–	kJ.kmol⁻¹. K⁻¹	kg.m²/s². kmol.K	–	Btu. kmol⁻¹. R⁻¹	lbm.ft²/s². kmol.R
Velocity V	meter per second	m·s⁻¹	m/s	feet per second	ft·s⁻¹	ft/s
Viscosity μ	pascal second	Pa·s N·s·m⁻²	kg/m·s	Psi second	Psi s lbf·s·ft⁻²	lbm/ft·s
Volume V	cubic meter	m³	m³	cubic feet	ft³	ft³
Volumetric flow rate $\dot{V}$	cubic meter per second	m³·s⁻¹	m³/s	cubic feet per second	ft³·s⁻¹	ft³/s
Weight W	gravitational force	N	kg·m/s²	gravitational force	lbf	lbm·ft/s²
Work W	product of newton and meter	N·m	kg·m²/s²	foot pound-force	ft·lbf	lbm·ft²/s²

A.3   UNIT CONVERSIONS

(a) <u>General</u>

DIMENSION	METRIC	METRIC/ENGLISH
Acceleration	$1\ m/s^2 = 100\ cm/s^2$	$1\ m/s^2 = 3.2808\ ft/s^2$ $1\ ft/s^2 = 0.3048\ m/s2$
Area	$1\ m^2 = 10^4\ cm^2 = 10^6\ mm^2 = 10^{-6}\ km^2$	$1\ m^2 = 1550\ in^2 = 10.764\ ft^2$ $1\ ft^2 = 144\ in^2 = 0.09290304\ m^2$
Density	$1\ g/cm^3 = 1\ kg/L = 1000\ kg/m^3$	$1\ g/cm^3 = 62.428\ lbm/ft^3 = 0.036127\ lbm/in^3$ $1\ lbm/in^3 = 1728\ lbm/ft^3$ $1\ kg/in^3 = 0.062428\ lbm/ft^3$
Energy, Heat, Work, Internal energy, Enthalpy	$1\ kJ = 1000\ J = 1000\ N.m = 1\ kPa.m^3$ $1\ kJ/kg = 1000\ m^2/s^2$ $1\ kWh = 3600\ kJ$ $1\ cal = 4.1868\ J$ $1\ IT\ cal = 4.1868\ J$ $1\ Cal = 4.1868\ J$	$1\ kJ = 0.94782\ Btu$ $1\ Btu = 1.055056\ kJ = 5.40395$ paia.ft3 $= 778.169\ lbf.ft$ $1\ Btu/lbm = 25037\ ft^2/s^2 = 2.326\ kJ/kg$ $1\ kJ/kg = 0.430\ Btu/lbm$ $1\ kWh = 3412.14\ Btu$ $1\ therm = 10^5\ Btu = 1.055 \times 10^5\ kJ$ (natural gas)
Force	$1\ N = 1\ kg.m/s^2 = 10^5\ dyne$ $1\ kgf = 9.80665\ N$	$1\ N = 0.22481\ lbf$ $1\ lbf = 32.174\ lbm{\cdot}ft/s^2 = 4.44822\ N$
Heat Flux	$1\ W/cm^2 = 10^4\ W/m^2$	$1\ W/m^2 = 0.3171\ Btu/h{\cdot}ft^2$
Heat Transfer Coefficient	$1\ W/m^2{\cdot}{}^\circ C = 1\ W/m^2{\cdot}K$	$1\ W/m^2{\cdot}{}^\circ C = 0.17612\ Btu/h{\cdot}ft^2{\cdot}\ {}^\circ F$
Length	$1\ m = 100\ cm = 1000\ mm = 10^6\ \mu m$ $1\ km = 1000\ m$	$1\ m = 39.270\ in = 3.2808\ ft = 1.0926\ yd$ $1\ ft = 12\ in = 0.3048\ m$ $1\ mile = 5280\ ft = 1.6093\ km$ $1\ in = 2.54\ cm = 25.4\ mm$
Mass	$1\ kg = 1000\ g$ $1\ metric\ ton = 1000\ kg$	$1\ kg = 2.2046226\ lbm$ $1\ bm = 0.45359237\ kg$ $1\ ouce = 28.3495\ g$ $1\ slug = 32.174\ lbm = 14.5939\ kg$ $1\ short\ ton = 2000\ lbm = 907.1847\ kg$

(*Continued*)

(Continued)

DIMENSION	METRIC	METRIC/ENGLISH
Power, Heat transfer rate	1 W = 1 J/s 1 kW = 1000 W = 1.341 hp 1 hp = 745.7 W	1 kW = 3412.14 Btu/h = 737.56 lbf·ft/s 1 hp = 550 ibf·ft/s = 0.7068 Btu/s = 42.41 Btu/min = 2544.5 Btu/h = 0.7457 kW 1 boiler hp = 33475 Btu/h 1 Btu/h = 1.055056 kJ/h 1 ton of refrigeration = 200 Btu/min
Pressure, Pressure head	1 Pa = 1 N/m^2 1 kPa = 10^3 Pa = 10^{-3} MPa 1 atm = 101.325 kPa = 1.01325 bars = 760 mm Hg@0°C = 1.03323 kgf/cm^2 1 mm Hg = 0.1333kPa	1 Pa = 1.4504x10^{-4} psia = 0.020886 lbf/ft^2 1 psi = 144 lbf/ft^2 = 6.894757 kPa 1 atm = 14.696 psia = 29.92 in Hg@30°F 1 in Hg = 3.387 kPa = 13.60 in H$_2$O
Specific heat	1 kJ/kg.°C = 1 kJ/kg.K = 1 J/g.C	1 Btu/lbm.°F = 4.1868 kJ/gk.°C 1 Btu/lbmol.R = 4.1868 kJ/kmol.K 1 kJ/kg.°C = 0.23885 Btu/lbm.°F = 0.23885 Btu/lbm.R
Specific volume	1 m^3/kg = 1000 L/kg = 1000 cm^3/g	1 m^3/kg = 16.02 ft^3/lbm 1ft^3/lbm = 0.062428 m^3/kg
Temperature	$T(K) = T(°C) + 273.15$ $\Delta T(K) = \Delta T(°C)$	$T(R) = T(°F) + 459.67 = 1.8 \times T(K)$ $T(°F) = 1.8 \times T(°C) + 32$ $\Delta T(°F) = \Delta T(R) = 1.8 \times \Delta T(K)$
Thermal conductivity	1 W/m.°C = 1 W/m.K	1 W/m.°C = o.57782 Btu/h.ft.°F
Thermal resistance	1 °C/W = 1 kW	1 kW = 0.52750 °F.h/Btu
Velocity	1 m/s = 3.60 km/h	1 m/s = 3.2808 ft/s = 2.237 mi/h 1 mi/h = 1.46667 ft/s 1 mi/h = 1.6093 km/h
Viscosity, dynamic	1 kg/m.s = 1 N.s/m2 = 1 Pa.s = 10 poise	1 kg/m.s = 2419.1 lbm/ft.h = 0.020886 lbf·s/ft^2 = 0.67197 lbm/ft.s

(Continued)

DIMENSION	METRIC	METRIC/ENGLISH
Viscosity, Kinematic	$0 \text{ m}^2/\text{s} = 10^4 \text{ cm}^2/\text{s}$ 1 stoke = $1 \text{ cm}^2/\text{s} = 10^{-4} \text{ m}^2/\text{s}$	$1 \text{ m}^2/\text{s} = 10.764 \text{ ft}^2/\text{s} = 3.875 \times 10^4$ ft^2/h
Volume	$1 \text{ m}^3 = 1000 \text{ L} = 10^6 \text{ cm}^3$ (cc)	$1 \text{ m}^3 = 6.1024 \times 10^4 \text{ in}^3 = 10^6$ $\text{cm}^3 = 35.315 \text{ ft}^3 = 264.17$ gal (U.S.) 1 gal (U.S.) = $231 \text{ in}^3 = 3.7854 \text{ L}$ 1 gal (U.S.) = 128 fl ounce 1 fl ounce = 29.5735 cm^3 = 0.0295735 L
Volume flow rate	$1 \text{ m}^3/\text{s} = 60,000 \text{ L/min} = 10^6 \text{ cm}^3/\text{s}$	$1 \text{ m}^3/\text{s} = 15,850$ gal/min (gpm) = $35.315 \text{ ft}^3/\text{s}$ = 2118.9 ft3 /min (CFM)

(b) Unit Conversion Factor (UCF)

Dimension	Unit Conversion	Unit Conversion Factor
Length	meter to millimeters	1 m = 1,000 mm
	meter to centimeters	1 m = 100 cm
	foot to inches	1 ft = 12 in.
	mile to feet	1 mi = 5,280 ft
	meter to feet	1 m = 3.2808 ft
	inch to millimeters	1 in. = 25.4 mm
Acceleration	m/s^2 to cm/s^2	$1 \text{ m/s}^2 = 100 \text{ cm/s}^2$
	m/s^2 to ft/s^2	1 m/s^2 to 3.2808 ft/s^2
Area	square meter to square millimeters	$1 \text{ m}^2 = 1,000,000 \text{ mm}^2$
	square meter to square feet	$1 \text{ m}^2 = 10.764 \text{ ft}^2$
	square foot to square inches	$1 \text{ ft}^2 = 144 \text{ in}^2$
	acre to square feet	1 acre = $43,560 \text{ ft}^2$
	acre to square meters	1 acre= $4,046 \text{ m}^2$
	square mile to acres	$1 \text{ mi}^2 = 640$ acres
Force	Newton to kilogram and meter per square second	$1 \text{ N} = 1 \text{ kg·m/s}^2$
	pound-force to pound-mass and feet per square second	$1 \text{ lbf} = 32.174 \text{ lbm·ft/s}^2$
	Newton to pound-force	1 N = 0.2248 lbf
	kilogram force to Newton	1 kgf = 9.8067 N
Heat	calorie to joules	1 cal = 4.184 J
	kilojoule to joules	1 kJ = 1,000 J
	Joule to Newtown and meter	1 J = 1 N·m
	kilojoule to British temperature unit	1 kJ = 0.9478 Btu

(Continued)

(Continued)

Dimension	Unit Conversion	Unit Conversion Factor
	British temperature unit per pound to kilojoule per kilogram	1 Btu/lbm = 2.356 kJ/kg
	therm to British temperature unit	1 therm = 100,000 Btu
Mass	kilogram to grams	1 kg = 1,000 g
	metric tonnage to kilograms	1 metric ton = 1,000 kg
	kilogram to pound-mass	1 kg = 2.2046 lbm
	slug to pound-mass	1 slug = 32.174 lbm
	slug to kilograms	1 slug = 14.5939 kg
Power	watt to J/s	1 watt = 1 J/s
	kilowatt to watts	1 kW = 1,000 W
	kilowatt to horsepower	1 kW = 1.341 hp
	kilowatt to Btu/h	1 kW = 3,412.14 Btu/h
	kilowatt to lbf·ft/s	1 kW = 737.56 lbf·ft/s
	horsepower to Btu/h	1 hp = 2,544.5 Btu/h
Pressure	atm to kilopascals	1 atm = 101.325 kPa
	atm to pressure per square inch	1 atm = 14.696 psi
	kilopascal to pascals	1 kPa = 1,000 Pa
	pressure per square inch to pascals	1 psi = 6.8948 kPa
	pascal to N/m^2	1 Pa = 1 N/m^2
	atm to millimeter mercury	1 atm = 760 mm Hg @ 0 °C
	atm to inch mercury	1 atm = 29.92 in Hg @ 30 °C
	inch mercury to kilopascals	1 in. Hg = 3.387 kPa
Time	minute to seconds	1 min = 60 s
	hour to minutes	1 h = 60 min
	Day to hours	1 d = 24 h
	week to days	1 wk = 7 d
	year to moths	1 a = 12 mo
	year to days	1 a = 365 d
Velocity	meter per second to kilometers per hour	1 m/s = 3.60 km/h
	meter per second to feet per second	1 m/s = 3.2808 ft/s
	mile per hour to kilometers per hour	1 mi/h = 1.6093 km/h
	knot to meters per second	1 knot = 0.5144 m/s
	mile per hour to feet per second	1 mi/h = 1.4667 ft/s
	knot to feet per second	1 knot = 1.6878 ft/s
Volume	liter to gallons	1 L = 0.2642 gal
	cubic foot to gallons	1 ft^3 = 7.4805 gal
	gallon to cubic meters	1 gal = 0.003485 m^3
	cubic foot cubic inches	1 ft^3 = 1,728 in^3
Work	kilowatt and hour to kilojoules	1 kWh = 3,600 kJ
	kilowatt and hour to British temperature unit	1 kWh = 3,412.14 Btu
	British temperature unit to pound-force and feet	1 Btu = 778.169 lbf·ft

(*Continued*)

(c) <u>Unity Conversion Ratio (UCR)</u>

Dimension	Unit Conversion	Unit Conversion Factor	Unity Conversion Ratio
Length	meter to millimeters	1 m = 1,000 mm	$\dfrac{1,000\text{mm}}{1\text{m}}$
	meter to centimeters	1 m = 100 cm	$\dfrac{100\text{cm}}{1\text{ m}}$
	foot to inches	1 ft = 12 in	$\dfrac{12\text{ in.}}{1\text{ ft}}$
	mile to feet	1 mi = 5,280 ft	$\dfrac{5,280\text{ ft}}{1\text{ mile}}$
	meter to feet	1 m = 3.2808 ft	$\dfrac{3.2808\text{ ft}}{1\text{ m}}$
	inch to millimeters	1 in. = 25.4 mm	$\dfrac{25.4\text{ mm}}{1\text{ in.}}$
Acceleration	m/s² to cm/s²	1 m/s² = 100 cm/s²	$\dfrac{100\,\frac{\text{cm}}{\text{s}^2}}{1\,\frac{\text{m}}{\text{s}^2}}$
	m/s² to ft/s²	1 m/s² = 3.2808 ft/s²	$\dfrac{3.2808\,\frac{\text{ft}}{\text{s}^2}}{1\,\frac{\text{m}}{\text{s}^2}}$
Area	square meter to square millimeters	1 m² = 1,000,000 mm²	$\dfrac{1,000,000\text{ mm}^2}{1\text{ m}^2}$
	square meter to square feet	1 m² = 10.764 ft²	$\dfrac{10.764\text{ft}^2}{1\text{m}^2}$
	square foot to square inches	1 ft² = 144 in.²	$\dfrac{144\text{ in.}^2}{\text{ft}^2}$
	acre to square feet	1 acre = 43,560 ft²	$\dfrac{43,560\text{ ft}^2}{1\text{ acre}}$
	acre to square meters	1 acre= 4,046 m²	$\dfrac{4,046\text{ m}^2}{1\text{ acre}}$
	square mile to acres	1 mi² = 640 acres	$\dfrac{640\text{ acres}}{1\text{mi}^2}$
Force	Newton to kilogram and meter per square second	1 N = 1 kg.m/s²	$\dfrac{1\text{ kg.}\frac{\text{m}}{\text{s}^2}}{1\text{N}}$
	pound-force to pound-mass and feet per square second	1 lbf = 32.174 lbm.ft/s²	$\dfrac{32.174\,\text{lbm.}\frac{\text{ft}}{\text{s}^2}}{1\,\text{lbf}}$
	Newton to pound-force	1 N = 0.2248 lbf	$\dfrac{0.2248\text{ lbf}}{1\text{N}}$
	kilogram-force to Newton	1 kgf = 9.8067 N	$\dfrac{9.8067\text{ N}}{1\text{ kgf}}$

(*Continued*)

(*Continued*)

Dimension	Unit Conversion	Unit Conversion Factor	Unity Conversion Ratio
Heat	calorie to joules	1 cal = 4.184 J	$\dfrac{4.184 \text{ J}}{1 \text{ cal}}$
	kilojoule to joules	1 kJ = 1,000 J	$\dfrac{1,000 \text{ J}}{1 \text{ kJ}}$
	Joule to Newtown and meter	1 J = 1 N.m	$\dfrac{1 \text{ N.m}}{1 \text{ J}}$
	kilojoule to British temperature unit	1 kJ = 0.9478 Btu	$\dfrac{0.9478 \text{ Btu}}{1 \text{ kJ}}$
	British temperature unit per pound to kilojoule per kilogram	1 Btu/lbm = 2.356 kJ/kg	$\dfrac{2.356 \frac{\text{kJ}}{\text{kg}}}{1 \frac{\text{Btu}}{\text{lbm}}}$
	therm to British temperature unit	1 therm = 100,000 Btu	$\dfrac{100,000 \text{ Btu}}{1 \text{ therm}}$
Mass	kilogram to grams	1 kg = 1,000 g	$\dfrac{1,000 \text{ g}}{1 \text{ kg}}$
	metric tonnage to kilograms	1 ton = 1,000 kg	$\dfrac{1,000 \text{ kg}}{1 \text{ ton}}$
	kilogram to pound-mass	1 kg = 2.2046 lbm	$\dfrac{2.2046 \text{ lbm}}{1 \text{ kg}}$
	slug to pound-mass	1 slug = 32.174 lbm	$\dfrac{32.174 \text{ lbm}}{1 \text{ slug}}$
	slug to kilograms	1 slug = 14.5939 kg	$\dfrac{14.5939 \text{ kg}}{1 \text{ slug}}$
Power	watt to J/s	1 watt = 1 J/s	$\dfrac{1 \frac{\text{J}}{\text{s}}}{1 \text{ watt}}$
	kilowatt to watts	1 kW = 1,000 W	$\dfrac{1,000 \text{ W}}{1 \text{ kW}}$
	kilowatt to horsepower	1 kW = 1.341 hp	$\dfrac{1.341 \text{ hp}}{1 \text{ kW}}$
	kilowatt to Btu/h	1 kW = 3,412.14 Btu/h	$\dfrac{3,412.14 \frac{\text{Btu}}{\text{h}}}{1 \text{ kW}}$
	kilowatt to lbf.ft/s	1 kW = 737.56 lbf.ft/s	$\dfrac{737.56 \text{ lbf.} \frac{\text{ft}}{\text{s}}}{1 \text{ kW}}$
	horsepower to Btu/h	1 hp = 2,544.5 Btu/h	$\dfrac{2,544.5 \frac{\text{Btu}}{\text{h}}}{1 \text{ hp}}$
Pressure	atm to kilopascals	1 atm = 101.325 kPa	$\dfrac{101.325 \text{ kPa}}{1 \text{ atm}}$
	atm to pressure per square inch	1 atm = 14.696 psi	$\dfrac{14.696 \text{ psi}}{1 \text{ atm}}$

(Continued)

Dimension	Unit Conversion	Unit Conversion Factor	Unity Conversion Ratio
	kilopascal to pascals	1 kPa = 1,000 Pa	$\dfrac{1,000 \text{ Pa}}{1 \text{ kPa}}$
	pressure per square inch to pascals	1 psi = 6.8948 kPa	$\dfrac{6.8948 \text{ kPa}}{1 \text{ psi}}$
	pascal to N/m²	1 Pa = 1 N/m²	$\dfrac{1 \text{ N}/\text{m2}}{1 \text{ Pa}}$
	atm to millimeter mercury	1 atm = 760 mm Hg @ 0 °C	$\dfrac{760 \text{ mm Hg}}{1 \text{ atm}}$
	atm to inch mercury	1 atm = 29.92 in Hg @ 30 °C	$\dfrac{29.92 \text{ in Hg}}{1 \text{ atm}}$
	inch mercury to kilopascals	1 in. Hg = 3.387 kPa	$\dfrac{3.387 \text{ kPa}}{1 \text{ in}}$
Time	minute to seconds	1 min = 60 s	$\dfrac{60 \text{ s}}{1 \text{ min}}$
	hour to minutes	hour = 60 min	$\dfrac{60 \text{ min}}{\text{hour}}$
	day to hours	1 d = 24 h	$\dfrac{24 \text{ h}}{1 \text{ d}}$
	week to days	1 wk = 7 d	$\dfrac{7 \text{ d}}{1 \text{ wk}}$
	year to moths	1 a = 12 mo	$\dfrac{12 \text{ mo}}{1 \text{ a}}$
	year to days	1 a = 365 d	$\dfrac{365 \text{ d}}{1 \text{ a}}$
Velocity	meter per second to kilometers per hour	1 m/s = 3.60 km/h	$\dfrac{3.60 \frac{\text{km}}{\text{h}}}{1 \frac{\text{m}}{\text{s}}}$
	meter per second to feet per second	1 m/s = 3.2808 ft/s	$\dfrac{3.2808 \text{ ft}/\text{s}}{1 \frac{\text{m}}{\text{s}}}$
	mile per hour to kilometers per hour	1 mi/h = 1.6093 km/h	$\dfrac{1.6093 \frac{\text{km}}{\text{h}}}{1 \frac{\text{mi}}{\text{h}}}$
	knot to meters per second	1 knot = 0.5144 m/s	$\dfrac{0.5144 \frac{\text{ft}}{\text{s}}}{1 \text{ knot}}$
	mile per hour to feet per second	1 mi/h = 1.4667 ft/s	$\dfrac{1.4667 \frac{\text{ft}}{\text{s}}}{1 \frac{\text{mi}}{\text{h}}}$
	knot to feet per second	1 knot = 1.6878 ft/s	$\dfrac{1.6878 \frac{\text{ft}}{\text{s}}}{1 \text{ knot}}$
Volume	liter to gallons	1 L = 0.2642 gal	$\dfrac{0.2642 \text{ gal}}{1 \text{ L}}$

(Continued)

(*Continued*)

Dimension	Unit Conversion	Unit Conversion Factor	Unity Conversion Ratio
	cubic foot to gallons	$1 \text{ ft}^3 = 7.4805 \text{ gal}$	$\dfrac{7.4805 \text{ gal}}{1 \text{ ft}^3}$
	gallon to cubic meters	$1 \text{ gal} = 0.003485 \text{ m}^3$	$\dfrac{0.003485 \text{ m}^3}{1 \text{ gal}}$
	cubic foot cubic inches	$1 \text{ ft}^3 = 1{,}728 \text{ in}^3$	$\dfrac{1{,}728 \text{ ft}^3}{1 \text{ in}^3}$
Work	kilowatt and hour to kilojoules	$1 \text{ kWh} = 3{,}600 \text{ kJ}$	$\dfrac{3{,}600 \text{ kJ}}{1 \text{ kWh}}$
	kilowatt and hour to British temperature unit	$1 \text{ kWh} = 3{,}412.14 \text{ Btu}$	$\dfrac{3{,}412.14 \text{ Btu}}{1 \text{ kWh}}$
	British temperature unit to pound-force and feet	$1 \text{ Btu} = 778.169 \text{ lbf·ft}$	$\dfrac{778.169 \text{ lbf·ft}}{1 \text{ Btu}}$
	pound-force and foot to lbm and square feet per square second	$1 \text{ lbf·ft} = 32.174 \text{ lbm·ft}^2/\text{s}^2$	$\dfrac{32.174 \text{ lbm·}\dfrac{\text{ft}^2}{\text{s}^2}}{1 \text{ lbf·ft}}$

A.4   DECIMAL AND THOUSANDS SEPARATORS

Locale	Number
Canadian (English and French)	5 382 697 415,000
Danish	5 382 697 415,000
Finnish	5 382 697 415,000
French	5 382 697 415,000
German	5 382 697.415,000
Italian	5.382.697.415,000
Norwegian	5.382.697.415,000
Spanish	5.382.697.415,000
Swedish	5 382 697 415,000
GB-English	5,382,697,415.000
US-English	5,382,697,415.000
Thai	5,382,697,415.000

Data files containing locale-specific formats are frequently misinterpreted when transferred to a system in a different locale. For example, a file containing numbers in a French format is not useful to a U.K.-specific program.

A.5   NON-SI UNITS ACCEPTED FOR USE WITH THE SI

Quantity	Name of unit	Symbol for unit	Value in SI units
Time	minute	min	1 min = 60 s
	hour	h	1 h = 60 min = 3,600 s
	day	d	1 d = 24 h = 86,400 s
Length	astronomical unit [a]	au	1 au = 149,597,870,700 m
Plane and phase angle	degree	°	$1° = (\pi/180)$ rad
	minute	′	$1' = (1/60)° = (\pi/10,800)$ rad
	second [b]	″	$1'' = (1/60)'' = (\pi/648,00)$ rad
Area	hectare [c]	ha	1 ha = 1 hm^2 = 10^4 m^2
Volume	liter [d]	l, L	1 l = 1 L = 1 dm^3 = 10^3 cm^3 = 10^{-3} m^3
Mass	tonne [e]	t	1 t = 10^3 kg
	dalton [f]	Da	1 Da = 1.660,539,066,60(50) × 10^{-27} kg
Energy	electronvolt [g]	eV	1 eV = 1.602,176,634 × 10^{-19} J
Logarithmic	neper [h]	Np	see text
Ratio quantities	bel [h]	B	
	decibel [h]	dB	

[a] As decided at the XXVIII General Assembly of the International Astronomical Union (Resolution B2, 2012).

[b] For some applications such as in astronomy, small angles are measured in arcseconds (i.e. seconds of plan angle), denoted as or, or milliarcseconds, microarcseconds, and picoarcseconds, denoted mas, µas, and pas, respectively, where arcsecond is an alternative name for second of plane angle.

[c] The unit hectare and its symbol, ha, were adopted by the CIPM in 1879 (PV, 1879, 41). The hectare is used to express land area.

[d] The liter and the symbol, lowercase l, were adopted by the CIPM in 1879 (PV, 1879, 41). The alternative symbol, capital L, was adopted by the 16th CGPM (1979, Resolution 6; CR, 101 and *Metrologia*, 1980, 16, 56–57) to avoid the risk of confusion between the letter l (el) and the numeral 1 (one).

[e] The tonne and its symbol, t, were adopted by the CIPM in 1879 (PV, 1879, 41). This unit is sometimes referred to as a "metric ton" in some English-speaking countries.

[f] The dalton (Da) and the unified atomic mass unit (u) are alternative names (and symbols) for the same unit, equal to 1/12 of the mass of a free carbon 12 atom, at rest and in its ground state. This value of the dalton is the value recommended in the CODATA 2018 adjustment.

[g] The electron volt is the kinetic energy acquired by an electron in passing through a potential difference of one volt in a vacuum. The electronvolt is often combined with the SI prefixes.

[h] When using these units, it is important that the nature of the quantity be specified and that any reference value used be specified.

A.6 DECISION OF THE CGPM AND THE CIPM

Since the SI is not a static convention but evolves following developments in the sciences of measurement, some decisions have been abrogated or modified; others have been clarified by additions. The decisions of the CGPM and CIPM are listed in strict chronological order, from 1889 to 2022, to preserve the continuity with which they were taken.

Decisions relating to the establishment of the SI

9th CGPM, 1948:	decision to establish the SI
10th CGPM, 1954:	decision on the first six base units
CIPM 1956:	decision to adopt the name "Système International d'Unités"
11th CGPM, 1960:	confirms the name and the abbreviation "SI", names prefixes from tera to pico, establishes the supplementary units rad and sr, lists some derived units
CIPM, 1969:	declarations concerning base, supplementary derived and coherent units, and the use of prefixes
CIPM, 2001:	"SI units" and "units of the SI"
23rd CGPM, 2007:	possible redefinition of certain base units of the International System of Units, the SI
24th CGPM, 2011:	possible future revision of the International System of Units, the SI
25th CGPM, 2014:	future revision of the International System of Units, the SI
26th CGPM, 2018:	revision of the International System of Units, the SI (to enter into force on 20 May 2019)

Decisions relating to the fundamental units of the SI

Length

1st CGPM, 1889:	sanction of the prototype meter
7th CGPM, 1927:	definition and use of the prototype meter
10th CGPM, 1954:	meter adopted as a base unit
11th CGPM, 1960:	redefinition of the meter in terms of krypton 86 radiation
15th CGPM, 1975:	recommends value for the speed of light
17th CGPM, 1983:	redefinition of the meter using the speed of light realization of the definition of the meter
CIPM, 2002:	specifies the rules for the practical realization of the definition of the meter
CIPM, 2003:	revision of the list of recommended radiations
CIPM, 2005:	revision of the list of recommended radiations

CIPM, 2007:	revision of the list of recommended radiations
23th CGPM, 2007:	revision of the *mise en pratique* of the definition of the meter and development of new optical frequency standards
CIPM, 2009:	updates to the list of standard frequencies
24th CGPM, 2011:	possible future revision of the International System of Units, the SI
24th CGPM, 2011:	revision of the *mise en pratique* of the definition of the meter and development of new optical frequency standards
CIPM, 2013:	updates to the list of standard frequencies
26th CGPM, 2018:	revision of the International System of Units, the SI (to enter into force on 20 May 2019)

Mass

1st CGPM, 1889:	sanction of the prototype kilogram
3rd CGPM, 1901:	declaration on distinguishing mass and weight and on the conventional value of g
10th CGPM, 1954:	kilogram adopted as a base unit
CIPM, 1967:	declaration on applying prefixes to the gram
21st CGPM, 1999:	future redefinition of the kilogram
23rd CGPM, 2007:	possible redefinition of certain base units of the International System of Units (SI)
24th CGPM, 2011:	possible future revision of the International System of Units, the SI
25th CGPM, 2014:	future revision of the International System of Units, the SI
26th CGPM, 2018:	revision of the International System of Units, the SI (to enter into force on 20 May 2019)

Time

10th CGPM, 1954:	second adopted as a base unit
CIPM, 1956:	definition of the second as a fraction of the tropical year 1900
11th CGPM, 1960:	ratifies the CIPM 1956 definition of the second
CIPM, 1964:	declares the caesium 133 hyperfine transition to be the recommended standard
12th CGPM, 1964:	empowers CIPM to investigate atomic and molecular frequency standards
13th CGPM, 1967/68:	defines the second in terms of the caesium transition
CCDS, 1970:	defines International Atomic Time, TAI
14th CGPM, 1971:	requests the CIPM to define and establish International Atomic Time, TAI
15th CGPM, 1975:	endorses the use of Coordinated Universal Time, UTC
CIPM, 2006:	secondary representations of the second

23rd CGPM, 2007:   on the revision of the *mise en pratique* of the definition of the meter and the development of new optical frequency standards

CIPM, 2009:   updates to the list of standard frequencies 189

24th CGPM, 2011:   possible future revision of the International System of Units, the SI

24th CGPM, 2011:   revision of the *mise en pratique* of the meter and the development of new optical frequency standards

CIPM, 2013:   updates to the list of standard frequencies

CIPM, 2015:   updates to the list of standard frequencies

26th CGPM, 2018:   revision of the International System of Units, the SI (to enter into force on 20 May 2019)

Electrical units

CIPM, 1946:   definitions of coherent electrical units in the meter-kilogram-second (MKS) system of units (to enter into force on 1 January 1948)

10th CGPM, 1954:   ampere adopted as a base unit

14th CGPM, 1971:   adopts the name siemens, symbol S, for electrical conductance

18th CGPM, 1987:   forthcoming adjustment to the representations of the volt and of the ohm

CIPM, 1988:   conventional value of the Josephson constant defined (to enter into force on 1 January 1990)

CIPM, 1988:   conventional value of the von Klitzing constant defined (to enter into force on 1 January 1990)

23rd CGPM, 2007:   possible redefinition of certain base units of the International System of Units (SI)

24th CGPM, 2011:   possible future revision of the International System of Units, the SI

25th CGPM, 2014:   future revision of the International System of Units, the SI

26th CGPM, 2018:   revision of the International System of Units, the SI (to enter into force on 20 May 2019)

Thermodynamic temperature

9th CGPM, 1948:   adopts the triple point of water as the thermodynamic reference point, adopts the zero of Celsius temperature to be 0.01 degrees below the triple point

CIPM, 1948:   adopts the name degree Celsius for the Celsius temperature scale

10th CGPM, 1954:   defines thermodynamic temperature such that the triple point of water is 273.16 degrees Kelvin exactly, defines standard atmosphere

10th CGPM, 1954:	degree Kelvin adopted as a base unit
13th CGPM, 1967/68:	decides formal definition of the kelvin, symbol K
CIPM, 1989:	the International Temperature Scale of 1990, ITS-90
CIPM, 2005:	note added to the definition of the kelvin concerning the isotopic composition of water
23rd CGPM, 2007:	clarification of the definition of the kelvin, unit of thermodynamic temperature
23rd CGPM, 2007:	possible redefinition of certain base units of the International System of Units (SI)
24th CGPM, 2011:	possible future revision of the International System of Units, the SI
25th CGPM, 2014:	future revision of the International System of Units, the SI
26th CGPM, 2018:	revision of the International System of Units, the SI (to enter into force on 20 May 2019)

Amount of substance

14th CGPM, 1971:	definition of the mole, symbol mol, as a seventh base unit, and rules for its use
21st CGPM, 1999:	adopts the special name katal, kat
23rd CGPM, 2007:	on the possible redefinition of certain base units of the International System of Units (SI)
24th CGPM, 2011:	possible future revision of the International System of Units, the SI
25th CGPM, 2014:	future revision of the International System of Units, the SI
26th CGPM, 2018:	revision of the International System of Units, the SI (to enter into force on 20 May 2019)

Luminous intensity

CIPM, 1946:	definition of photometric units, new candle and new lumen (to enter into force on 1 January 1948)
10th CGPM, 1954:	candela adopted as a base unit
13th CGPM, 1967/68:	defines the candela, symbol cd, in terms of a black body radiator
16th CGPM, 1979:	redefines the candela in terms of monochromatic radiation
24th CGPM, 2011:	possible future revision of the International System of Units, the SI
26th CGPM, 2018:	revision of the International System of Units, the SI (to enter into force on 20 May 2019)

Decisions relating to SI-derived and supplementary units

SI-derived units

12th CGPM, 1964:	accepts the continued use of the curie as a non-SI unit
13th CGPM, 1967/68:	lists some examples of derived units
15th CGPM, 1975:	adopts the special names becquerel, Bq, and gray, Gy
16th CGPM, 1979:	adopts the special name sievert, Sv
CIPM, 1984:	decides to clarify the relationship between absorbed dose (SI unit gray) and dose equivalent (SI unit sievert)
CIPM, 2002:	modifies the relationship between absorbed dose and dose equivalent

Supplementary units

CIPM, 1980:	decides to interpret supplementary units as dimensionless derived units
20th CGPM, 1995:	decides to abrogate the class of supplementary units, and confirms the CIPM interpretation that they are dimensionless derived units

A.7 DIMENSIONLESS NUMBERS NAMED AFTER PEOPLE

Dimensionless Number	Named After	Facts	Expression
Archimedes number (Ar)	Archimedes of Syracuse (287–212 BC)	Greek mathematician and physicist.	$\dfrac{gL^3\rho_s(\rho_s-\rho)}{\mu^2}$
Biot number (Ai)	Biot, Jean-Baptiste (1774–1862)	French mathematician and physicist.	$\dfrac{hL}{k}$
Darcy friction factor (f_D)	Darcy, Henry P. G. (1803–1858)	French engineer	$\dfrac{8\tau_w}{\rho V^2}$
Eckert number (Ec)	Eckert, Ernst R. G. (1904–2004)	Austrian American engineer and scientist	$\dfrac{V^2}{c_p T}$
Euler number (Eu)	Euler, Leonhard (1707–1783)	Swiss mathematician and physicist	$\dfrac{\Delta P}{\rho V^2}$
Fanning friction factor (C_f)	Fanning, John T. (1837–1911)	American architect and hydraulic engineer	$\dfrac{2\tau_w}{\rho V^2}$
Fourier number (Fo)	Fourier, Jean B. J. (1837–1830)	French mathematician and physicist	$\dfrac{\alpha t}{L^2}$
Froude number (Fr)	Froude, William (1810–1879)	English engineer	$\dfrac{V}{\sqrt{gL}}$
Grashof number (Gr)	Grashof, Franz (1826–1893)	German engineer and professor	$\dfrac{g\beta\Delta T L^3\rho^2}{\mu^2}$
Jakob number (Ja)	Jakob, Max (1879–1955)	German physicist	$\dfrac{c_p\left(T-T_{sat}\right)}{h_{fg}}$
Knudsen Number (Kn)	Knudsen, Martin (1871–1949)	Danish physicist	$\dfrac{\lambda}{L}$
Lewis number (Le)	Lewis, Warren K. (1882–1975)	American engineer and professor	$\dfrac{k}{\rho c_p D_{AB}}$
Mach number (Ma)	Mach, Ernst (1838–1916)	Czech physicist and philosopher	$\dfrac{V}{c}$
Nusselt number (Nu)	Nusselt, Wilhelm (1882–1957)	German Engineer	$\dfrac{Lh}{k}$
Peclet number (Pe)	Peclet, Jean C. E. (1793–1857)	French physicist	$\dfrac{\rho L V c_p}{k}$
Prandtl number (Pr)	Prandtl, Ludwig (1875–1953)	German physicist and aerospace scientist	$\dfrac{\mu c_p}{k}$
Rayleigh number (Ra)	Rayleigh, John W. S. (1842–1919)	British mathematician and physicist	$\dfrac{g\beta\Delta T L^3\rho^2 c_p}{k\mu}$
Reynolds number (Re)	Reynolds, Osborne (1842–1912)	British engineer	$\dfrac{\rho V L}{\mu}$

(Continued)

Dimensionless Number	Named After	Facts	Expression
Richardson number (Ri)	Richardson, Lewis (1881–1953)	English mathematician and physicist	$\dfrac{g\beta\Delta L^5}{\rho V^2}$
Schmidt number (Sc)	Schmidt, Ernst (1892–1975)	German scientist	$\dfrac{\mu}{\rho D_{AB}}$
Sherwood number (Sh)	Sherwood, Thomas K. (1903–1976)	American chemical engineer	$\dfrac{VL}{D_{AB}}$
Stanton number (St)	Stanton, Thomas E. (1865–1931)	British mechanical engineer	$\dfrac{h}{\rho c_p V}$
Stokes number (StK)	Stokes, George G. (1819–1903)	Irish scientist	$\dfrac{\rho_p D_p^2 V}{18\mu L}$
Strouhal number (St)	Strouhal, Vincenz (1850–1922)	Czech physicist	$\dfrac{fL}{V}$
Ursell number (U)	Ursell, Fritz J. (1923–2012)	British mathematician	$\dfrac{H\lambda^2}{3}$
Weber number (We)	Weber, Moritz (1871–1951)	German professor	$\dfrac{\rho V^2 L}{\sigma_s}$

A.8 FE MECHANICAL EXAM AND PE MECHANICAL–THERMAL AND FLUID SYSTEMS EXAM

(a) <u>FE Mechanical Exam Topics</u>

Fundamentals of Engineering (FE)
MECHANICAL CBT Exam Specifications
(Effective Beginning with the July 2020 Examinations)

- The FE exam is a computer-based test (CBT). It is a closed-book test with an electronic reference.
- The FE exam uses both the International System of Units (SI) and the U.S. Customary System (USCS).
- Examinees have 6 hours to complete the exam, which contains 110 questions. The 6-hour time also includes a tutorial and an optional scheduled break.

Knowledge	Number of Questions
1 Mathematics	6–9
A Analytic geometry	
B Calculus (e.g., differential, integral, single-variable, multivariable)	
C Ordinary differential equations (e.g., homogeneous, nonhomogeneous, Laplace transforms)	
D Linear algebra (e.g., matrix operations, vector analysis)	
E Numerical methods (e.g., approximations, precision limits, error propagation, Taylor's series, Newton's method)	
F Algorithm and logic development (e.g., flowcharts, pseudocode)	
2 Probability and Statistics	4–6
A Probability distributions (e.g., normal, binomial, empirical, discrete, continuous)	
B Measures of central tendencies and dispersions (e.g., mean, mode, standard deviation, confidence intervals)	
C Expected value (weighted average) in decision making	
D Regression (linear, multiple), curve fitting, and goodness of fit (e.g., correlation coefficient, least squares)	
3 Ethics and Professional Practice	4–6
A Codes of ethics (e.g., NCEES Model Law, professional and technical societies, ethical and legal considerations)	
B Public health, safety, and welfare	
C Intellectual property (e.g., copyright, trade secrets, patents, trademarks)	
D Societal considerations (e.g., economic, sustainability, life-cycle analysis, environmental)	
4 Engineering Economics	4–6
A Time value of money (e.g., equivalence, present worth, equivalent annual worth, future worth, rate of return, annuities)	
B Cost types and breakdowns (e.g., fixed, variable, incremental, average, sunk)	
C Economic analyses (e.g., cost-benefit, break-even, minimum cost, overhead, life cycle)	

(Continued)

		Knowledge	Number of Questions
5		Electricity and Magnetism	5–8
	A	Electrical fundamentals (e.g., charge, current, voltage, resistance, power, energy, magnetic flux)	
	B	DC circuit analysis (e.g., Kirchhoff's laws, Ohm's law, series, parallel)	
	C	AC circuit analysis (e.g., resistors, capacitors, inductors)	
	D	Motors and generators	
6		Statics	9–14
	A	Resultants of force systems	
	B	Concurrent force systems	
	C	Equilibrium of rigid bodies	
	D	Frames and trusses	
	E	Centroids and moments of inertia	
	F	Static friction	
7		Dynamics, Kinematics, and Vibrations	10–15
	A	Kinematics of particles	
	B	Kinetic friction	
	C	Newton's second law for particles	
	D	Work-energy of particles	
	E	Impulse-momentum of particles	
	F	Kinematics of rigid bodies	
	G	Kinematics of mechanisms	
	H	Newton's second law for rigid bodies	
	I	Work-energy of rigid bodies	
	J	Impulse-momentum of rigid bodies	
	K	Free and forced vibrations	
8		Mechanics of Materials	9–14
	A	Shear and moment diagrams	
	B	Stress transformations and Mohr's circle	
	C	Stress and strain caused by axial loads	
	D	Stress and strain caused by bending loads	
	E	Stress and strain caused by torsional loads	
	F	Stress and strain caused by shear	
	G	Stress and strain caused by temperature changes	
	H	Combined loading	
	I	Deformations	
	J	Column buckling	
	K	Statically indeterminate systems	
9		Material Properties and Processing	7–11
	A	Properties (e.g., chemical, electrical, mechanical, physical, thermal)	
	B	Stress-strain diagrams	
	C	Ferrous metals	
	D	Nonferrous metals	
	E	Engineered materials (e.g., composites, polymers)	

(Continued)

(Continued)

		Knowledge	Number of Questions
	F	Manufacturing processes	
	G	Phase diagrams, phase transformation, and heat treating	
	H	Materials selection	
	I	Corrosion mechanisms and control	
	J	Failure mechanisms (e.g., thermal failure, fatigue, fracture, creep)	
10		Fluid Mechanics	10–15
	A	Fluid properties	
	B	Fluid statics	
	C	Energy, impulse, and momentum	
	D	Internal flow	
	E	External flow	
	F	Compressible flow (e.g., Mach number, isentropic flow relationships, normal shock)	
	G	Power and efficiency	
	H	Performance curves	
	I	Scaling laws for fans, pumps, and compressors	
11		Thermodynamics	10–15
	A	Properties of ideal gases and pure substances	
	B	Energy transfers	
	C	Laws of thermodynamics	
	D	Processes	
	E	Performance of components	
	F	Power cycles	
	G	Refrigeration and heat pump cycles	
	H	Nonreacting mixtures of gases	
	I	Psychrometrics	
	J	Heating, ventilation, and air-conditioning (HVAC) processes	
	K	Combustion and combustion products	
12		Heat Transfer	7–11
	A	Conduction	
	B	Convection	
	C	Radiation	
	D	Transient processes	
	E	Heat exchangers	
13		Measurements, Instrumentation, and Controls	5–8
	A	Sensors and transducers	
	B	Control systems (e.g., feedback, block diagrams)	
	C	Dynamic system response	
	D	Measurement uncertainty (e.g., error propagation, accuracy, precision, significant figures)	

(Continued)

	Knowledge	Number of Questions
14	Mechanical Design and Analysis	10–15
A	Stress analysis of machine elements	
B	Failure theories and analysis	
C	Deformation and stiffness	
D	Springs	
E	Pressure vessels and piping	
F	Bearings	
G	Power screws	
H	Power transmission	
I	Joining methods (e.g., welding, adhesives, mechanical fasteners)	
J	Manufacturability (e.g., limits, fits)	
K	Quality and reliability	
L	Components (e.g., hydraulic, pneumatic, electromechanical)	
M	Engineering drawing interpretations and geometric dimensioning and tolerancing (GD&T)	

The NCEES splits the Fundamentals of Engineering (FE) exam into seven disciplines as shown. The disciplines are designed to cover six major engineering fields with an additional "Other Disciplines" category as a catchall for engineers who do not fall into the major disciplines.

- Chemical
- Civil
- Electrical and Computer
- Environmental
- Industrial and Systems
- Mechanical
- Other Disciplines

For detailed topics and categories of each discipline exam, visit www.prepfe.com/fe-exams/disciplines.

(b) <u>PE Mechanical Exam Topics</u>

Principles and Practice of Engineering Examination (PE)
MECHANICAL—THERMAL AND FLUID SYSTEMS CBT Exam
Specifications
(Effective Beginning April 2020)

- The exam topics have not changed since April 2017 when they were originally published.

- The exam is computer-based. It is closed-book exam with an electronic reference.
- Examinees have 9 hours to complete the exam, which contains 80 questions. The 9-hour time includes a tutorial and an optional scheduled break. Examinees work all questions.
- The exam uses both the International System of Units (SI) and the U.S. Customary System (USCS).
- The exam is developed with questions that require a variety of approaches and methodologies, including design, analysis, and application.
- The knowledge areas specified as examples of kinds of knowledge are not exclusive or exhaustive categories.

Knowledge			Number of Questions
I	Principles		28–44
	A	Basic Engineering Practice	5–8
		1 Engineering terms, symbols, and technical drawings	
		2 Economic analysis	
		3 Units and conversions	
	B	Fluid Mechanics	5–8
		1 Fluid properties (e.g., density, viscosity)	
		2 Compressible flow (e.g., Mach number, nozzles, diffusers)	
		3 Incompressible flow (e.g., friction factor, Reynolds number, lift, drag)	
	C	Heat Transfer Principles (e.g., convection, conduction, radiation)	5–8
	D	Mass Balance Principles (e.g., evaporation, dehumidification, mixing)	4–6
	E	Thermodynamics	5–8
		1 Thermodynamic properties (e.g., enthalpy, entropy)	
		2 Thermodynamic cycles (e.g., Combined, Brayton, Rankine)	
		3 Energy balances (e.g., 1st and 2nd laws)	
		4 Combustion (e.g., stoichiometrics, efficiency)	
	F	Supportive Knowledge	4–6
		1 Pipe system analysis (e.g., pipe stress, pipe supports, hoop stress)	
		2 Joints (e.g., welded, bolted, threaded)	
		3 Psychrometrics (e.g., dew point, relative humidity)	
		4 Codes and standards	
II	Hydraulic and Fluid Applications		21–33
	A	Hydraulic and Fluid Equipment	13–21
		1 Pumps and fans (e.g., cavitation, curves, power, series, parallel)	
		2 Compressors (e.g., dynamic head, power, efficiency)	
		3 Pressure vessels (e.g., design factors, materials, pressure relief)	
		4 Control valves (e.g., flow characteristics, sizing)	
		5 Actuators (e.g., hydraulic, pneumatic)	
		6 Connections (e.g., fittings, tubing)	
	B	Distribution Systems (e.g., pipe flow)	8–12

(Continued)

		Knowledge	Number of Questions
III		Energy/Power System Applications	21–33
	A	Energy/Power Equipment	7–11
		1 Turbines (e.g., steam, gas)	
		2 Boilers and steam generators (e.g., heat rate, efficiency)	
		3 Internal combustion engines (e.g., compression ratio, BMEP)	
		4 Heat exchangers (e.g., shell and tube, feedwater heaters)	
		5 Cooling towers (e.g., approach, drift, blowdown)	
		6 Condensers (e.g., surface area, materials)	
	B	Cooling/Heating (e.g., capacity, loads, cycles)	5–8
	C	Energy Recovery (e.g., waste heat, storage)	5–8
	D	Combined Cycles (e.g., components, efficiency)	4–6

The NCEES PE exam is offered in 27 disciplines. The exam includes multiple-choice questions as well as alternative item types (AITs). Three disciplines covered on the Mechanical exam are

- Mechanical: HVAC and Refrigeration,
- Mechanical: Machine Design and Materials, and
- Mechanical: Thermal and Fluid Systems.

See other disciplines and learn more about exam-specific requirements and information on the format of the exam and the topics, Visit https://ncees.org/exams/pe-exam.

Bibliography

A. C. Palmer, *Dimensional Analysis and Intelligent Experimentation*. World Scientific Publishing, Hackensack, NJ, 2008.

Ambler Thompson, Barry N. Taylor, *Guide for the Use of the International System of Units (SI)*. NIST Special Publication 811, NIST, 2008.

American Society of Heating, Refrigerating and Air-Conditioning Engineers (ASHRAE), *Handbook of Fundamentals*. The American Society of Heating, Refrigerating and Air-Conditioning Engineers, 2017.

ANSI/ASME Y14.5M-1994 (R1999), *Dimensioning and Tolerances—Mathematical Definitions of Principles*. American Society of Mechanical Engineers (ASME), New York, 1994.

ASME Y14.38–1999, *Abbreviations and Acronyms*. American Society of Mechanical Engineers (ASME), New York, 1999.

British Standards Institution, *Specification for SI Units and Recommendations for the Use of Their Multiples and of Certain Other Units*. British Standards Institution, 1993.

Bruce R. Munson, Donald F. Young, Theodore H. Okiishi, *Fundamentals of Fluid Mechanics*. John Wiley & Sons, Inc., New York, 1990.

Carl F. Gauss, "Intensitas vis Magneticae Terrestris ad Mensuram Absolutam Revocata". *Commentationes Societatis Regiae Scientiarum Gottingensis Recentiores*, 8: 3–44, 1832 (Susan P. Johnson, "The Intensity of the Earth's Magnetic Force Reduced to Absolute Measurement". Translated from the German, July 1995).

Cecil Jensen, Jay D. Helsel, Dennis R. Short, *Engineering Drawing and Design*. McGraw Hill Education, seventh edition, New York, 2002.

D. C. Ipsen, *Units, Dimensions, and Dimensionless Numbers*. McGraw-Hill, New York, 1960.

Don S. Lemons, *A Student's Guide to Dimensional Analysis (Student's Guides)*. Cambridge University Press, student edition, March 2017.

E. Buckingham, "On Physically Similar Systems: Illustrations of the Use of Dimensional Equations". *Physical Review*, 4: 345–376, 1914.

E. S. Taylor, *Dimensional Analysis for Engineers*. Clarendon Press, Oxford, 1974.

Elisa A. Artusa, *SI (Metric) Handbook*. Technical Memorandum 109197, March 1994.

Eugene A. Avallone, Theodore Baumeister, Ali Sadegh, Lionel S. Marks, *Marks' Standard Handbook for Mechanical Engineers*. McGraw Hill, eleventh edition, New York, USA, 2006.

Frank M. White, *Fluid Mechanics*. McGraw Hill Education, eighth edition, New York, 2016.

Frederick E. Giesecke, Alva Mitchell, Henry C. Spencer, Ivan L. Hill, *Technical Drawing*. Pearson, thirteenth edition, New Jersey and Ohio, January 2008.

G. P. Craig, *Clinical Calculations Made Easy: Solving Problems Using Dimensional Analysis*. Lippincott Williams and Silkins, fourth edition, Baltimore, MD, 2008.

IEEE, *ASTM SI10–16, IEEE/ASTM SI 10 American National Standard for Metric Practice*. IEEE, New York, July 2016.

International Bureau of Weights and Measures (2019), *The International System of Units (SI)*. BIPM, ninth edition, V2, December 1, 2022.

International Bureau of Weights and Measures (BIPM): Joint Committee for Guides in Metrology, *International Vocabulary of Metrology—Basic and General Concepts and Associated Terms (VIM)*. BIPM, third edition, 2012. Retrieved March 28, 2015.

ISO 1000, *SI Units and Recommendations for the Use of Their Multiples and of Certain Other Units*. ISO, 1992.

ISO, *ISO 80000-1:2022 Quantities and Units Part 1: General*. ISO, second edition, 2022.

J. Kunes, *Dimensionless Physical Quantities in Science and Engineering*. Elsevier, New York, 2012.

J. L. Meriam, L. G. Kraige, *Engineering Mechanics—Dynamics*. John Wiley & Sons, Inc., seventh edition, Danvers, MA, 2012.

J. L. Meriam, L. G. Kraige, J. N. Bolton, *Engineering Mechanics—Dynamics*. John Wiley & Sons, Inc., ninth edition, New York, 2018.

J. L. Meriam, L. G. Kraige, J. N. Bolton, *Engineering Mechanics—Statics*. John Wiley & Sons, Inc., ninth edition, Danvers, MA, 2018.

John, J. Bertin, Russell M. Commings, *Aerodynamics for Engineers*, Pearson Education, Inc., fifth edition, Upper Saddle River, NJ, 2009.

Kenneth S. Butcher, Linda D. Crown, Elizabeth J. Gentry, Carol Hockert, *The International System of Units (SI)—Conversion Factors for General Use*. NIST Special Publication 1038, NIST, May 2006.

L. E. Barbrow, L. V. Judson, *Weights and Measures of the United States*. National Bureau of Standards Special Publication, 1976.

Michael R. Lindeburg, *Core Engineering Concepts for Students and Professionals*. Professional Publications Inc., Belmont, CA, 2010.

Michael R. Lindeburg, *FE Mechanical Review Manual*. PPI, Kaplan Company, 2014.

Michael R. Lindeburg, *FE Other Disciplines Review Manual*. PPI, Kaplan Company, 2014.

Michael R. Lindeburg, *FE Review Manual*. Professional Publications Inc., 2004.

Michael R. Lindeburg, *Mechanical Engineering Reference Manual for the PE Exam*. Professional Publications Inc., twelfth edition, Belmont, CA, 2006.

National Council of Examiners for Engineering and Surveying (NCEES), *FE Reference Handbook 10.0.1*. NCEES, 2020.

National Institute of Standards and Technology (NIST), *Guide for the Use of the International System of Units (SI)*. NIST Special Publication 811, NIST, 2008.

National Institute of Standards and Technology (NIST), *The International System of Units (SI)*. NIST Special Publication 330, NIST, 2008.

P. W. Bridgman, *Dimensional Analysis*. Hale University Press, New Haven, CT, 1922, revised edition, 1963.

Physical Measurement Laboratory, Special Publication 811. NIST, Created January 2016 and updated November 2019.

Resolution 10 of the 22nd CGPM, *Symbol for the Decimal Marker*. The 22nd General Conference, CGPM, 2003.

Robert A. Granger, Fluid Mechanics. CBS College Publishing, New York, 1985.

Russ Rowlet, *Units: CGS and MKS*. University of North Carolina, Chapel Hill, May 2018. Retrieved May 2021.

SAE Metric Advisory Committee, *Rules for SAE Use of SI (Metric) Units*. Technical Standards Board Standard, May 1999.

The World Factbook, *Weights and Measures*. Central Intelligence Agency, Washington, DC, April 22, 2021.

Yongjian Gu, *Heating and Cooling of Air Through Coils*. CRC Press, Boca Raton, 2024.

Yunus A. Cengel, Afshin J. Ghajar, *Heat and Mass Transfer—Fundamentals and Applications*. McGraw Hill Education, sixth edition, New York, 2020.

Yunus A. Cengel, John M. Cimbala, *Fluid Mechanics—Fundamentals and Applications*. McGraw Hill Education, fourth edition, New York, 2018.

Yunus A. Cengel, Michael A. Boles, *Thermodynamics—An Engineering Approach*. McGraw Hill Education, ninth edition, New York, 2018.

Index

Printed in the United States
by Baker & Taylor Publisher Services